Microprocessor Architecture, Programming, and Systems Featuring the 8085

William A. Routt

THOMSON

DELMAR LEARNING

Australia • Canada • Mexico • Singapore • Spain • United Kingdom • United States

THOMSON

★

DELMAR LEARNING

™

Microprocessor Architecture, Programming, and Systems Featuring the 8085
William Routt

Vice President, Technology and Trades ABU:
David Garza

Director of Learning Solutions:
Sandy Clark

Managing Editor:
Larry Main

Senior Product Manager:
Michelle Ruelos Cannistraci

Senior Acquisitions Editor:
Stephen Helba

Marketing Director:
Deborah S. Yarnell

Marketing Manager:
Guy Baskaran

Marketing Coordinator:
Patti Garrison

Editorial Assistant:
Dawn Daugherty

Director of Production:
Patty Stephan

Production Manager:
Andrew Crouth

Content Project Manager:
Benj Gleeksman

Technology Project Manager:
Linda Verde

Library of Congress Cataloging-in-Publication Data

Routt, William A.
  Microprocessor architecture, programming, and systems featuring the 8085/by William A. Routt.
    p. cm.
  Includes index.
  ISBN 1-4180-3241-7
  1. Intel 8085 (Microprocessor)
  2. Microprocessors.   I. Title.

QA76.8.I2912.R68 2007
004.1'65—dc22

2006022765

NOTICE TO THE READER

Publisher does not warrant or guarantee any of the products described herein or perform any independent analysis in connection with any of the product information contained herein. Publisher does not assume, and expressly disclaims, any obligation to obtain and include information other than that provided to it by the manufacturer.

The reader is expressly warned to consider and adopt all safety precautions that might be indicated by the activities herein and to avoid all potential hazards. By following the instructions contained herein, the reader willingly assumes all risks in connection with such instructions.

The publisher makes no representation or warranties of any kind, including but not limited to, the warranties of fitness for particular purpose or merchantability, nor are any such representations implied with respect to the material set forth herein, and the publisher takes no responsibility with respect to such material. The publisher shall not be liable for any special, consequential, or exemplary damages resulting, in whole or part, from the readers' use of, or reliance upon, this material.

Contents

Chapter 5   Programming the 8085 – Basic Commands   73

Chapter 6   Programming the 8085 – Advanced Techniques   95

Chapter 7   General-Purpose Support Chips   119

Chapter 10 Comparison of the 8085 to Other Microprocessors   189

Preface

This textbook is intended to be used as a first resource on microprocessors at the two-year associate degree level or the four-year technical degree level, and includes enough material for a full-length semester course on microprocessors. The lecture information contained within presents questions at the end of each chapter which can serve as homework assignments, as well as problems comparable to real-world problems that students may encounter when working with microprocessors.

The laboratory experiments in most of the chapters are intended to be run in a lab setting using an 8085 Microprocessor Trainer. Each requires two to three hours to execute.

This textbook is also meant to serve as a good overall review of microprocessors for any practicing engineer or technician.

Conceptual Approach

This textbook begins by introducing the student to a generic microprocessor, both from a hardware and a software level. It presents a simplified assembly language that allows the student to learn what assembly language is all about, utilizing a very reduced and simplified instruction set. During my years of teaching microprocessor courses to associate degree students, I've often found that one of the hardest concepts for the hardware student is learning how to write assembly language programs. However, by first introducing a very simple generic instruction set, there is much less for the student to learn in the area of instructions, and they can thus concentrate on how to actually write an assembly language program.

I also introduce flowcharting as well as pseudo-code in this book to provide students with two of the most commonly used tools in program design. These tools are more often used in higher-level programming, but they can work just as well at the assembly language level.

This textbook concentrates on the 8085A 8-bit microprocessor since I feel that once a student understands how an 8-bit microprocessor works, they can easily learn how to use a 16- or 32-bit microprocessor, all of which work in essentially the same manner except that the latter two have wider buses and require a few more instructions.

I chose the 8085A because it is an Intel microprocessor and because Intel has such a huge presence in the microprocessor market today. In addition, the 8085A has many trainers and simulators available, given it's been around for so long and is currently used at so many schools.

I also spend an entire chapter on the 8051 microcontroller since I've always felt that microcontrollers as a group do not usually get enough coverage in the typical microprocessor textbook.

Assumed Prerequisites

For this book, I assume that the student has an understanding of digital electronics as well as knowledge of the hexadecimal and binary numbering systems.

The Organization of the Text

The textbook is organized into six areas:

- A definition of a microprocessor (Chapter 1)
- A hardware and software description of a generic microprocessor (Chapters 2 and 3)
- A detailed hardware and software description of the 8085A microprocessor (Chapters 4, 5, and 6)
- A description of 8085A support chips and system applications (Chapters 7 and 8)
- A detailed hardware and software description of the 8051 microcontroller (Chapter 9)
- A comparison of the 8085A to other microprocessors (Chapter 10)

Chapter 1 discusses what a microprocessor is and how it compares to a microcontroller and a microcomputer.

Chapter 2 covers the hardware architecture of a generic microprocessor. It also explores the ALU, the system bus, memory types, and I/O methods of programming.

Chapter 3 discusses how to program a generic microprocessor, covers the classifications of instructions, shows a generic instruction set, explains how to write an assembly language program, shows how to create a flowchart, and explores what pseudo-code is.

Chapter 4 describes the hardware architecture of the 8085A microprocessor and discusses both memory interfacing and I/O interfacing to the 8085A.

Chapter 5 covers the basic commands for programming the 8085A microprocessor.

Chapter 6 presents some of the more advanced programming techniques used in programming the 8085A such as looping, counting, the stack, subroutines, and conditional call and return instructions.

Chapter 7 discusses five of the most common support chips used with the 8085A.: the 8255 PPI, the 8254 PIT, the 8259 PIC, the 8237 DMAC, and the 8155 Memory IO/M/T chip.

Chapter 8 describes how to interface several devices to an 8085A, such as a matrix display, a matrix keyboard, an A/D converter, and a D/A converter. Single-board computers are also covered.

Chapter 9 offers a complete hardware and software description of the Intel 8051 microcontroller.

Chapter 10 compares the 8085A to several other microprocessors, such as Motorola's 6800, Motorola's 68000, Intel's 8088, Intel's 80186 and 80286, Intel's 80386 and 80486, and Intel's Pentium.

Appendix I presents a complete listing of the 8085A instruction set.

Appendix II is composed of the Intel specification sheets for the 8085A microprocessor.

Appendix III contains the Intel specification sheets of the 8051A microcontroller.

Each chapter starts off with objectives for that chapter followed by the keywords (and their definitions) used in that chapter. At the end of each chapter is a bulleted summary of the material covered.

Questions are posted at the end of each chapter addressing all of the content presented. Also, Chapters 3 through 9 include problems covering the material learned. These questions and problems make for good homework assignments. Chapters 3 through 8 offer laboratory experiments which are intended to be run in a lab setting using an 8085 Microprocessor Trainer. Chapter 9 has two labs for use with an 8051 Microcontroller Trainer.

The Recommended Sequence of Study

The recommended sequence of study for a first course in microprocessors is to cover Chapter 1 through Chapter 8 sequentially. If time permits, Chapter 9 should be covered next, followed by Chapter 10. The inclusion of microcontrollers is obviously up to the instructor, so some may choose to jump from Chapter 8 directly to Chapter 10.

Supplemental Material

An Instructor's Manual is provided, which contains answers to all of the chapter questions and problems. This manual also includes answers to the lab questions that appear at the end of seven of the chapters in the book.

In addition, a PowerPoint presentation of the entire book is provided in the supplemental material. The Instructor's Manual and PowerPoints are available on CD (ISBN: 1418032425).

Acknowledgments

I wish to thank my wonderful wife, Trish, who was a veritable fountain of support, encouragement, and understanding during the year it took me to write this book.

I also wish to thank Delmar Publishing for giving me the opportunity to write a book on microprocessors that is tailored to how I currently teach my course on microprocessors.

In addition, I'd like to thank John Clevenger, a fellow instructor at Wake Technical Community College, who wrote the set of 8085 laboratory experiments we currently use at Wake Tech, and which were used as the basis for several lab experiments in this book.

I'd also like to express my appreciation for the entire editing staff at Delmar for all their help with this book. The author would also like to thank the following reviewers:

Alan Essenmacher, Henry Ford Community College, Dearborn, MI
Gerald Gambs, Pennsylvania Institute of Technology, Media, PA
Frank Lanzer, Anne Arundel Community College, Arnold, MD
Michael Pelletier, Northern Essex Community College, Salem, MA
Mukul V. Shirvaikar, University of Texas, Tyler, TX

About the Author

William A. Routt has a Bachelor's Degree in Electrical Engineering from Pennsylvania State University, and a Master's Degree in Electrical Engineering from Carnegie Institute of Technology. He was an Assistant Professor in the Engineering Technology Department at University of North Carolina at Charlotte, and is currently a department head of the Electronics Engineering Technology Department and the Computer Engineering Technology Department at Wake Technical Community College. He has been teaching at the four-year technical degree level and the two-year technical associate degree level for the past 14 years.

1

An Introduction to Microprocessors

Objectives:

Upon completion of this chapter, you should:

- Understand what the microprocessor is

- Know what a microprocessor-based system contains

- Understand assembly language programming versus high-level programming

- Comprehend how a microprocessor executes instructions

- Understand the definition of an embedded system

Key Terms:

- **Microprocessor**—A programmable, clocked, multipurpose, electronic device

- **Machine language**—Instructions that are in the binary form

- **Microcontroller**—A complete computer system on a single chip

- **Reprogrammable system**—A system that includes a general-purpose microprocessor that can be programmed for a variety of tasks

- **Embedded system**—A system where the microprocessor is programmed and then embedded in a final product that does not require any reprogramming of the microprocessor

- **Assembly language**—Low-level programming language that allows people to write programs at a mnemonic level instead of in binary form

- **Microcomputer**—A complete computer system built around a microprocessor

Introduction

The basis for many of our electronic systems today, the **microprocessor** is an integrated circuit and a programmable logic device which can be used to control processes and turn different devices on and off. It can also be employed as the data processing unit or computing unit of a computer, and as the CPU for a microcomputer.

The microprocessor runs on **machine language**, or instructions that are in the binary form. However, we write code for a microprocessor in assembly language—a low-level language—which uses mnemonics for the instructions, instead of straight binary. Low-level languages require the programmer to be much more familiar with the hardware architecture and the register layout of the microprocessor. High-level languages are those such as C, C++, Fortran, and Java.

1-1 The Microprocessor

The microprocessor is a programmable, multipurpose, clocked, electronic device that reads binary instructions from a memory component and then executes these instructions via its control unit. It can then output data to a memory component, or control signals to output devices. It can also accept inputs from a variety of input devices. Input and output devices are called peripherals.

A typical programmable machine is considered to be composed of four components: the microprocessor, memory, input, and output (as shown in Figure 1-1).

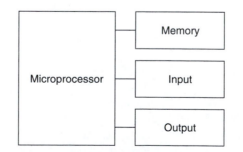

Figure 1-1 ■ A programmable machine

Such components are considered hardware, while a set of instructions written for the microprocessor to perform specific tasks is called a program (also known as software). The machine shown in Figure 1-1 can be programmed to perform a variety of tasks such as turning traffic lights on and off, solving mathematical functions, or controlling a home security system.

A microprocessor can be a CPU for a general-purpose computer and can also be embedded into a microcontroller, which is a microprocessor included in a complete controller system on one chip used for specific applications.

Typically, a CPU is composed of a control unit and an arithmetic and logic unit (ALU). Figure 1-2 shows a microprocessor as the CPU in a general-purpose computer.

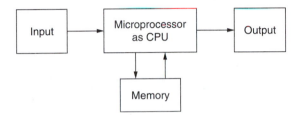

Figure 1-2 ■ A microprocessor as the CPU in a computer

A **microcontroller** is a complete computer system on a single chip. It includes a microprocessor, memory, and several peripheral components such as I/O ports, timers, and A/D converters. Microcontrollers are often used for specific applications. In such cases, they are preprogrammed. Figure 1-3 shows a block diagram for a microcontroller.

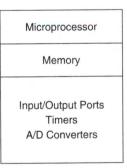

Figure 1-3 ■ A block diagram for a microcontroller

1-2 A Microprocessor-Based System

Microprocessor-based systems can be generally grouped into two categories: **reprogrammable systems** and **embedded systems.** Reprogrammable systems include general-purpose microprocessors that can be reprogrammed for a variety of tasks. The personal computer (PC) is a typical example of a reprogrammable system.

An embedded system, on the other hand, is one where the microprocessor is programmed and then included or embedded in a final product that does not require any reprogramming of the microprocessor. A copy machine is a typical example of an embedded system, as is the fuel-injection controller in an automobile. In these instances, the user is not allowed to reprogram the microprocessor.

A microcontroller (discussed later in detail in Chapter 9) is basically a microprocessor and its peripherals, and is often used in these embedded systems.

Figure 1-4 shows a simplified structure of a microprocessor-based system, where a bus is used as a communication path between the microprocessor, input/output, and memory.

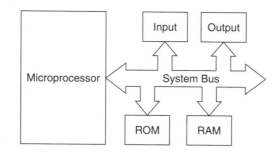

Figure 1-4 ■ A simplified microprocessor-based system

The entire system in Figure 1-4 is also called a microcomputer, which is different from a microprocessor. A microprocessor is one component of a microcomputer, which is a complete computer system, and only differs from a regular computer in that the CPU is a microprocessor. However, microcomputers and microprocessor-based systems can essentially be used interchangeably.

The Microprocessor

The microprocessor is a clock-driven semiconductor device that is manufactured using LSI or VLSI technology. In large computers, the CPU (central processing unit) performs the same operations that a microprocessor performs. It can execute instructions and perform arithmetic and logical operations. The microprocessor is on one chip and is composed of an arithmetic and logic unit (ALU), an array of registers, and a control unit—the same components found in a CPU.

The ALU part of the microprocessor performs the arithmetic operations such as addition and subtraction, as well as the logical operations such as ANDing, ORing, and Exclusive ORing.

The register array of the microprocessor contains various registers used primarily for the temporary storage of data during instruction execution and/or arithmetic/logic operations.

The control unit part of the microprocessor provides the necessary timing and control signals for all the operations in the microprocessor. It controls the flow of information between the microprocessor, memory, and the peripherals.

Memory

Memory stores binary information such as data or a program that acts as a set of instructions for the microprocessor. The data can be stored temporarily during operations by the microprocessor, or if it is the result of calculations performed by the microprocessor, it can be stored for a longer period of time. Two basic types of memory exist: read-only memory (ROM) and random access memory (RAM), which is read-write memory. ROM is used to store instructions or data that won't change since it can only be read, not altered. RAM is typically used for the temporary storage of data and user-written programs since it can be used on a read/write basis. It is also known as the user memory.

Input/Output

Input and output devices communicate with the outside world, and together are considered peripherals. Input devices enter data into the microprocessor. Typical devices of this sort are keyboards, switches, and analog-to-digital (A/D) devices. A/D devices convert an analog signal or level into a digital signal that can then be input into a microprocessor.

Output devices take data from the microprocessor and display it or present it to the outside world. Typical output devices include light emitting diodes (LEDs), printers, magnetic tape units, cathode ray tubes (CRTs), and monitors.

SYSTEM BUS

The system bus is a communication path between the microprocessor and its peripherals. It is a set of parallel wires, grouped together, to send parallel binary data to and from the microprocessor and the peripherals. The system bus is shared by the microprocessor and all of the peripherals. It therefore requires specific handshaking to facilitate this sharing.

1-3 Microprocessor Languages

Microprocessors operate on binary instructions that are formed into words. The format of these binary instructions differ for each microprocessor, but they all process them in basically the same way. The length of the word is defined as the number of bits the microprocessor recognizes in parallel and processes at one time. Typically, the word length for microprocessors goes from 4-bit words at the low end, all the way up to 64-bit words for high-performance microprocessors.

The term byte is used globally to mean a group of 8 bits. A 16-bit microprocessor is therefore said to have a word length of 2 bytes. The term nibble is used to represent 4 bits, thus two nibbles make up one byte.

Each microprocessor has its own unique set of instructions that is based on the specific hardware design of that microprocessor. However, there is a similarity between all microprocessors at the instruction set level. Only so many types of instructions are required for a microprocessor to execute, and once you've learned to program one microprocessor, it's very easy to program a different microprocessor. For example, there are only so many ways to abbreviate JUMP.

The microprocessor executes instructions that are in binary form, called machine instructions. However, because it's very hard for people to write programs (a set of instructions) in binary form, assembly language was created. **Assembly language** is low-level programming that allows people to write programs at a mnemonic level instead of in binary form. Assembly language is unique for every microprocessor, but each language uses mnemonics to represent every specific instruction that a particular microprocessor can recognize and execute.

High-level languages allow a person to write programs at a higher level than assembly language by using mnemonics that represent more complex operations than at the assembly language level. This allows a person to write programs that are machine-independent. A high-level language program can be written and then run on several different hardware platforms. When you write a program in a high level language, it can be compiled for running on different specific hardware machines. A compiler basically converts a high-level language into machine language for that specific processor. Therefore, a compiler is hardware-specific.

Figure 1-5 shows how a high-level BASIC program can be compiled for several different hardware microprocessors.

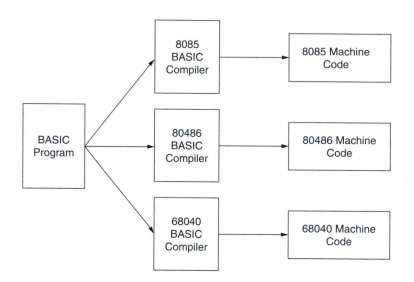

Figure 1-5 ■ High-level program compilers

As for assembly language, an assembler is used to convert the assembly language program into machine language. It is machine-specific, obviously, but so is each assembly language. For example, 8085 assembly language can only be assembled for, and run on, an 8085 microprocessor.

Figure 1-6 shows how a program written in 8085 assembly language can be assembled with an 8085 assembler, and converted into 8085 machine code that is then executed on an 8085 microprocessor.

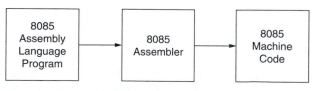

Figure 1-6 ■ The 8085 assembly language assembler

You can also hand-assemble an assembly language program for a microprocessor. It is much easier to use an assembler, which is a program that automatically generates the appropriate machine language from the assembly language, but you can convert the assembly language for a specific microprocessor to the appropriate machine language by hand. When hand-assembling, it is easier to write the machine language in hexadecimal. For example, the 8085 is an 8-bit microprocessor, so each instruction word can be represented by two hexadecimal digits. Each hex digit is 4 bits in binary.

To hand-assemble, all one needs is the table (Appendix I for the 8085 microprocessor) that gives the hex or binary equivalent for each instruction for that specific microprocessor. For example, the 8085 instruction INR A causes the 8085 microprocessor to increment the register A by one. The assembly language mnemonic is INR A. The hex equivalent for this instruction is 80H. This represents the binary 8-bit word of 1000 0000. When the control logic in the 8085 gets the binary word 1000 0000, it will increment register A by 1.

A program that has been hand-assembled can then be entered into the microprocessor via a hex keyboard. Many single-board computers, microcomputers, or microprocessor trainers have a hex keyboard for inputting hand-assembled programs. It should be noted that hand-assembling is okay for small programs, but much too tedious for larger programs. For larger programs, you need an assembler. An assembler is considered a cross-assembler if it runs on one microprocessor but generates machine code for another microprocessor. Many cross-assemblers exist, which are assemblers that (for example) translate 8085 assembly language mnemonics into 8085 machine code, but which run on some other microprocessor such as a Pentium-based PC. Programs that are converted to machine code by a cross-assembler often are then downloaded into the microprocessor via a serial link.

1-4 Microcomputers

The **microcomputer** is a microprocessor-based system, or rather it is a complete computer system that is built around a microprocessor. A microcomputer falls into the category of a reprogrammable microprocessor system. Early microcomputer systems were based on 8-bit microprocessors, while today the most common microcomputer is the personal computer (PC), which is based on 64-bit microprocessors like the Pentium or Motorola's 680XX.

The workstation, used in engineering and scientific applications, is another microcomputer that is basically a high-end PC. Indeed, as PCs get more and more powerful, the distinction between a workstation and a high-end PC becomes increasingly blurred.

Single-board microcomputers are complete microprocessor systems contained on a single printed circuit board. Single-board microcomputers are often used in both teaching labs as well as in industrial applications, and have their microprocessor, memory, and input/output on the same board.

Single-chip microcomputers are also known as microcontrollers. They contain the microprocessor, memory, and some input/output capability all on a single chip. Microcontrollers are typically used for specific applications such as controlling appliances, or are employed in industrial control applications.

Summary

- The microprocessor is a programmable, clocked, multipurpose, electronic device.
- The microprocessor executes instructions that are in binary form.
- A programmable machine has four components: microprocessor, memory, input, and output.
- A microprocessor can be a CPU for a general-purpose computer, and can also be embedded into a microcontroller.
- Microprocessor-based systems can be generally grouped into two classifications: reprogrammable systems and embedded systems.
- An embedded system is one where the microprocessor is programmed and then embedded in a final product that does not require any reprogramming of the microprocessor.
- The microprocessor is a clock-driven semiconductor device that is manufactured in LSI or VLSI technology.
- The microprocessor is composed of an arithmetic and logic unit (ALU), an array of registers, and a control unit.
- Memory stores binary information such as data or a program which acts as a set of instructions for the microprocessor and data.
- Input and output devices communicate with the outside world.
- The system bus is a communication path between the microprocessor and its peripherals.
- The word length for microprocessors goes from 4-bit words at the low end all the way up to 64-bit words for high-performance microprocessors.
- Each microprocessor has its own unique set of instructions that are based on the specific hardware design of that microprocessor.
- Assembly language is low-level programming that allows people to write programs at a mnemonic level instead of in binary form.
- High-level languages allow a person to write programs at a higher functional level than assembly language.
- A high-level language can be written and then run on several different hardware platforms.
- An assembler is used to convert the assembly language program into machine language. The assembler is machine-specific.
- You can hand-assemble an assembly language program for a microprocessor.
- Many single-board computers, microcomputers, or microprocessor trainers have a hex keyboard for inputting hand-assembled programs.
- The microcomputer is a microprocessor-based system, constituting a complete computer system built around a microprocessor.
- The most commonly known microcomputer is the personal computer (PC).

- A workstation is another microcomputer that is essentially a high-end PC.
- Single-board microcomputers are complete microprocessor systems contained on a single printed circuit board.

Questions

1. What is assembly language?
2. What is the difference between low-level languages and high-level languages?
3. How does a microprocessor perform functions?
4. What are the four basic components of a typical programmable machine?
5. What are the two general classifications of microprocessor-based systems?
6. What is the difference between a microprocessor and a microcomputer?
7. What does an ALU do?
8. Name the two basic types of memory.
9. List two typical input devices and two typical output devices.
10. Why does the system bus require handshaking?
11. Why are instructions specific to a particular microprocessor?
12. What is the function of a compiler?
13. What is the output of an assembler?
14. How do you hand-assemble a microprocessor program?
15. Name three examples of a microcomputer.

2

The Hardware Architecture of a General Microprocessor-Based System

Objectives:

Upon completion of this chapter, you should:

- Understand the hardware architecture of a generic microprocessor

- Learn about the use and components of the system bus

- Understand the general classifications of memory

- Learn about peripheral-mapped I/O and memory-mapped I/O

- Learn about typical Input/output devices

Key Terms:

- **System bus**—A bus that is composed of three separate buses: the address bus, the data bus, and the control bus

- **Address bus**—A one-way parallel bus that is used for sending the memory or I/O address

- **Data bus**—A two-way or bidirectional bus that is used to send data to and from the microprocessor

- **Control bus**—A one-way or unidirectional bus that is used for control signals sent from the microprocessor to memories or peripheral devices

- **Arithmetic and logic unit (ALU)**—Performs arithmetic operations such as add and subtract, as well as logic operations like AND, OR, and XOR

- **Accumulator**—A special-purpose register that is used in all of the arithmetic and logical operations

- **RAM (random access memory)**—Read/write memory consisting of two main types: static RAM (SRAM) and dynamic RAM (DRAM)

- **ROM (read-only memory)**—Memory that is read-only and nonvolatile

- **RAM disks**—Small removable devices that plug into the USB port and appear to a PC as a "removable drive"

- **Memory-mapped I/O**—A mapped I/O that does not use I/O instructions, but instead allows all of the memory instructions to be used on peripheral devices, giving these devices addresses within the memory address range

- **Peripheral-mapped I/O**—A mapped I/O in which the microprocessor can address any peripheral device by using I/O instructions and putting the address of the peripheral device out on the address bus

Introduction

The hardware architecture of a microprocessor-based system refers to the hardware components that make up the system, and how they are interconnected. The architecture itself refers to the internal logic design of the microprocessor.

Figure 2-1 shows the typical hardware architecture of a microprocessor-based system.

The **system bus** is composed of three separate buses: the **address bus**, the **data bus**, and the **control bus**. All typical microprocessor systems have a system bus made up of these three distinct buses. The microprocessor serves as the CPU (central processing unit), or MPU (microprocessing unit), of the system. The memory stores both the program that the microprocessor executes, as well as temporary data from computations or functions performed by the

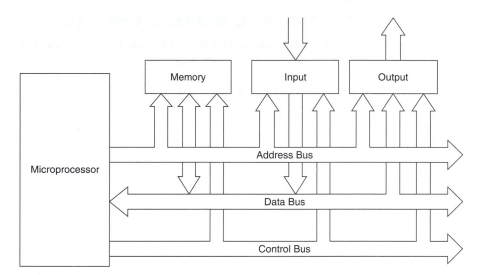

Figure 2-1 ■ The microprocessor-based system hardware architecture

microprocessor. The input device(s) allow data to be entered into the system, while the output device(s) display or send data out of the system.

2-1 The Microprocessor Architecture

A typical microprocessor contains an **arithmetic and logic unit (ALU)**, an **accumulator**, an instruction decoder, a control unit, registers, and an internal bus. Figure 2-2 shows the hardware architecture of a typical microprocessor.

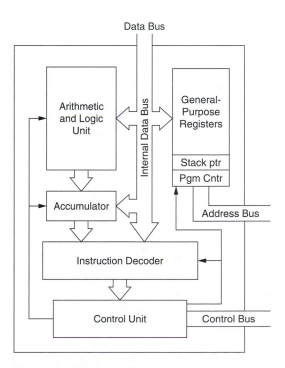

Figure 2-2 ■ The microprocessor hardware architecture

The Arithmetic and Logic Unit (ALU)

The arithmetic and logic unit (ALU) performs arithmetic operations such as add and subtract, and logic operations such as AND, OR, XOR, and so on. The ALU is directed as to what operations to perform by the control unit and uses the internal data bus to move data between the internal general-purpose registers. The accumulator is usually the register that holds the result of the arithmetic and logical operations.

When data is sent externally from the microprocessor or received from an external source, the data comes into, or goes out from, the accumulator via the data bus.

The Instruction Decoder

The instruction decoder decodes the instructions it receives over the internal data bus. The address of the next instruction to be executed is stored in the program counter, and when this address is sent out of the microprocessor to an external memory over the address bus, the resulting instruction (stored at that address in the memory) is returned over the data bus and is sent internally to the instruction decoder. The instruction decoder decodes the instruction at a binary level and sends the appropriate signals to the control unit.

The Control Unit

The control unit receives the information from the instruction decoder after it decodes the instruction as to what operations need to be performed. It then sends the binary signals to the appropriate internal sections of the microprocessor and dispatches any needed control signals over the control bus to external devices such as memories and input or output devices.

General-Purpose Registers

The general-purpose registers are used by the microprocessor when performing any of a number of operations. The size and number of registers vary depending upon which specific microprocessor it is. For example, the 8085 8-bit microprocessor has six 8-bit general-purpose registers. These registers are listed specifically in the assembly language instructions that compose the instruction set for the microprocessor. As a result, writing programs for a microprocessor is done at the register-level in that the programmer decides which registers to use for each instruction, and what data goes in which registers.

There are also special-purpose registers in all microprocessors. The accumulator is considered a special-purpose register because it is used in all ALU operations. The program counter is also a special-purpose register in that it always (and only) contains the memory address of the next instruction byte to be executed. The stack pointer is another special-purpose register and points to the top of a memory area called the stack that is used by the microprocessor. (The stack will be discussed more in later chapters.) One other special-purpose register is the flag register. This contains individual bits called flags that correspond to different error conditions or different computational conditions such as the accumulator is zero or non-zero, or a carry has been generated, or even or odd parity is present, and so on.

2-2 System Bus

The system bus is comprised of three distinct buses, which together are referred to as the system bus. These three components are the **address bus**, the **data bus**, and the **control bus**.

Figure 2-3 shows the three components of a typical system bus.

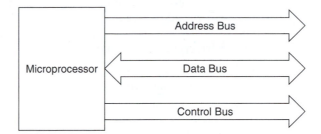

Figure 2-3 ■ A typical system bus

The Address Bus

The **address bus** is a parallel bus that is one way, from the microprocessor out to memories and peripheral devices. The width of the parallel bus determines how much memory that particular microprocessor can address. For example, the 80386 microprocessor has a 32-bit address bus, and can therefore address 2^{32}, or 4,294,967,296, memory locations. The 8085 has a 16-bit address bus, and can therefore address 2^{16}, or 65,536, memory locations. The size of the address bus is independent of the size of the microprocessor. A microprocessor such as the 8085 is called an 8-bit microprocessor, which refers to the size of the data bus, not the address bus.

The Data Bus

The **data bus** is a two-way or bidirectional bus and is used to send data to and from the microprocessor. Examples of data coming into the microprocessor could be data from a memory location that is an instruction to be executed, while an example of data going out from the microprocessor could be data going out from the microprocessor for display on some LEDs and/or a seven-segment display.

The data bus size determines the size of the microprocessor. For example, an 8-bit microprocessor has an 8-bit wide data bus, while the 80386, being a 32-bit microprocessor, has a 32-bit wide data bus. In addition, the size of the internal microprocessor general-purpose registers usually matches the size of the data bus. Thus, since the 8085 has 8-bit wide registers internally, its data bus is 8 bits wide. The size of the data bus matches the size of the internal registers so that all the bits on the bus can at one time come into, or go out of, one of the registers.

The Control Bus

The **control bus** is typically considered a one-way or unidirectional bus, sending signals out from the microprocessor to memories or peripheral devices. The size of the control bus depends upon the specific microprocessor. Typically, it contains signals that specify whether the microprocessor's operation is a read or a write, and whether it is a memory or input/output operation. Often, the control bus is not organized the same way as buses with adjacent pins like the address and data bus. Nor is it as well defined. Nevertheless, it is still considered a bus, and all microprocessors have a control bus. Other signals that may be included in the control bus signals are state signals and address latch enables. A microprocessor also has additional control signals—such as interrupt signals, acknowledgement signals, and hold signals—that usually aren't considered part of the control bus even though they help control the microprocessor-based system.

The Multiplexed Bus

Multiplexed buses occur in some microprocessors and deal with the sharing of the physical bus between the address bus and the data bus. It is usually done in the design of the microprocessor chip because of a shortage of available pins on the chip. For example, in the 8085 microprocessor, the same eight pins serve as the low-order 8 bits of the address bus, as well as the 8 bits of the data bus. Multiplexed buses require the use of the buses to be sequenced. When the low 8 bits of the multiplexed bus are being used for the full 16-bit address, it can't be employed as the data bus. In contrast, when the 8 bits of the multiplexed bus are being used as the data bus, they aren't available as the low-order half of the address bus. However, the multiplexed bus feature, though common, is not as desirable as separate buses because of how it slows down the operation of the microprocessor due to the sequencing of operations that must be done. Because of this, no overlapping of operations can occur with regards to the address bus and the data bus.

2-3 Memory

The memory in a microprocessor-based system is used to store the programs that operate the microprocessor, as well as temporary data employed by the microprocessor during computations and/or calculations. Memory is addressed via the address bus where the address of the memory location being read or written is sent out by the microprocessor. The memory recognizes that the address is located in its address block and determines via the control signals whether it is a read or write.

Memory can be classified into two major categories: main memory and secondary memory. Main memory is made up of **read-only memory (ROM)** and **random access memory (RAM)**. RAM is read/write memory, while ROM is obviously read-only. Microprocessor-based systems usually have both types of main memory. The program is often stored in ROM (for protection against being overwritten), and RAM is used for temporary storage of data. Smaller embedded systems, depending upon their function, may only have ROM and no RAM. However, most systems have both. It's also possible for a system to only have RAM and thus store the program there along with the temporary data.

Secondary storage memory devices include hard disks, floppy disks, CDs, tapes, and RAM disks. Microprocessor-based systems often have secondary memory devices such as hard drives or CDs based upon the function the system was designed for. The PC is a microprocessor-based system that has multiple secondary memory devices.

Figure 2-4 shows the major classifications of different types of memory.

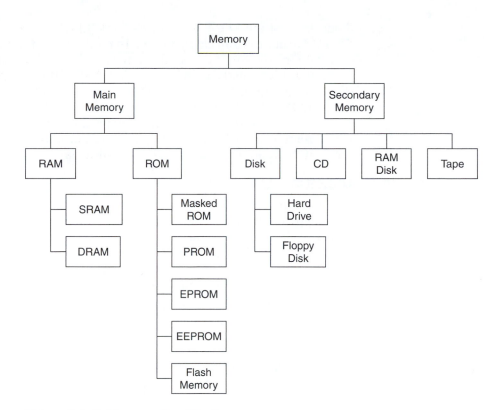

Figure 2-4 ■ Memory classifications

RAM

Random access memory (RAM) is read/write memory and is composed of two main types: static RAM (SRAM) and dynamic RAM (DRAM). SRAM uses an entire flip-flop to store each bit in memory. It is faster than DRAM, but is much less dense and more expensive. SRAM is often used in microprocessor systems as cache memory that's located on the microprocessor chip in order to speed up processor operations.

DRAM uses an electrical charge—as in a capacitor—to store each bit in memory. As a result, it's not as fast as SRAM, but is much denser and less expensive. All of the RAM memory used in microprocessor-based systems that is external to the microprocessor chip itself is always DRAM. When your PC lists that it has 256MB of RAM, this is DRAM. Another main characteristic of DRAM memories is that they require constant refreshing so they don't lose their information. Since each bit is basically a capacitive charge, it must be constantly refreshed—read out and then back in—so it isn't lost.

ROM

ROM memories are essentially classified by how they are programmed. Masked ROMs have the bit pattern permanently recorded on them. They cannot be changed, and it is an expensive procedure. Programmable ROMs are programmed via a PROM "burner" that permanently burns the bits into the diode array of the ROM. Erasable PROMs (EPROMs), on the other hand, use a PROM burner to store charges on FETs for the bits, but it isn't permanent. The entire memory can be erased using an ultraviolet light, and then reprogrammed. Electrically erasable PROMs (EEPROMs) are similar to EPROMS, except that the chip can be erased electrically. Flash memory is very similar to EEPROMs, but in this case the entire chip can be erased or only sections of it. Also, the newest flash memories can be erased with voltages as low as 1.8V. Flash memories are becoming popular in microprocessor-based systems since they can be reprogrammed or reloaded while still in the system.

Disks

Disks are secondary memory devices and can take the form of hard drives and floppy disks. Disks are rotating magnetic devices where the data is stored magnetically and is accessed sequentially per track, but the appropriate track is accessed randomly.. Thus, they are considered semi-sequential access devices. Though slower than RAM or ROM, they provide much more economical storage per bit, and are removable media. The floppy disk (much less used today) is removable media in all cases, and the hard drives can come in removable versions as well.

CDs

Compact disks (CDs) are very similar to magnetic disks except that they use light to store the data on a disk instead of magnetism. Very dense and removable, they are rapidly replacing floppy disks given their much larger storage capacity.

RAM Disks

Sometimes called thumbnail drives, jump drives, RAM sticks, or USB disks, **RAM disks** are not really disks at all. They are small, removable devices that plug into the USB port and appear to a PC as a "removable hard drive." Because of their convenience, they have already begun to replace floppy disks and CDs. Currently, they're available in 256MB and 512MB capacities and are continually increasing in size.

Tape

Magnetic tapes are still used in some instances, mainly for backup purposes, but they don't see much use except in larger computer systems where daily backup tapes are made. They are sequential devices.

2-4 Input/Output

Input and output devices allow the microprocessor-based system to communicate with the outside world, taking data in and then displaying or sending the data back out. There are two general methods by which microprocessor-based systems handle peripheral I/O devices: **peripheral-mapped I/O** and **memory-mapped I/O**.

Peripheral-Mapped I/O

Microprocessors contain a class of instructions called input/output (I/O) instructions, which refer to input and output devices only, and are used when the system uses peripheral-mapped I/O. In this method, the microprocessor can address any peripheral device by using I/O instructions and putting the address of the peripheral device out on the address bus. In peripheral-mapped I/O, the addresses used for the peripheral devices are outside the memory address spectrum. Therefore, the peripheral devices do not use up memory addresses, and there can be as many I/O addresses as there are memory addresses. Normally, you don't need nearly as many I/O addresses as you do memory addresses. This is an advantage of the system. The disadvantage is that you're limited to using only the I/O instructions and not the memory instructions, which are generally more powerful. Many microprocessor-based systems use peripheral-mapped I/O, particularly larger systems.

Memory-Mapped I/O

The memory-mapped I/O method does not use I/O instructions but instead allows all of the memory instructions to be used on peripheral devices, giving peripheral devices addresses within the memory address range. In this way, the peripheral devices "look" just like any memory location. The disadvantage of this method is that it uses up memory addresses from the memory address range for peripheral devices. The advantage, however, is that you get to use powerful and flexible memory instructions instead of the more restrictive I/O instructions.

Input Devices

Input devices are peripheral devices that inject data into the microprocessor system. A keyboard, for instance, is one of the most common input devices. It sends the ASCII code for each key that is pressed to the microprocessor. As with most input devices, the keyboard alerts the microprocessor that it has some data for it via interrupts, which are a way for peripheral devices to signal the microprocessor that they need service. Interrupts will be discussed in more detail in chapter 4.

Other input devices include the mouse, the light pen, the touch screen, the keypad, switches, and so on. All of which are devices that get information into the microprocessor in some fashion.

Output Devices

Output devices are peripheral devices that take data from the microprocessor and either display it, store it, or control external processes with it. The most common output device, at least in the realm of PCs, is the monitor, which displays information that the microprocessor sends to it. Monitors used to only employ cathode ray tubes (CRTs), but now flat screen LCD monitors are also available.

Other output devices include LEDs, seven-segment displays, relays, and so on.

Summary

- The hardware architecture of the microprocessor refers to its internal logic design.
- The system bus is composed of three separate buses: the address bus, the data bus, and the control bus.

- The microprocessor serves as the CPU (central processing unit), or the MPU (microprocessing unit) of the system.
- A typical microprocessor contains an arithmetic and logic unit (ALU), an accumulator, an instruction decoder, a control unit, registers, and an internal bus.
- The ALU performs arithmetic operations such as add and subtract, and logic operations like AND, OR, XOR, and so on.
- The instruction decoder decodes the instructions it receives over the internal data bus.
- The control unit receives the information from the instruction decoder after the instruction is decoded and then determines what operations need to be performed.
- The general-purpose registers are used by the microprocessor when performing any of a number of operations.
- The components of the system bus are the address bus, the data bus, and the control bus.
- The one-way address bus is a parallel bus that is used for sending the memory or I/O address to peripheral devices.
- The data bus is a two-way or bidirectional bus that is used to send data to and from the microprocessor.
- The control bus is a one-way or unidirectional bus that is used to send control signals from the microprocessor to memories or peripheral devices.
- Multiplexed busing is the sharing of the physical bus between the address bus and the data bus.
- The memory in a microprocessor-based system is used to store the programs that operate the microprocessor, as well as temporary data that is used by the microprocessor during computations and/or calculations.
- Memory can be classified into two major categories: main memory and secondary memory.
- Main memory is made up of read-only memory (ROM) and random access memory (RAM).
- Secondary storage memory devices include hard disks, floppy disks, CDs, tapes, and RAM disks.
- Random access memory (RAM) has two main types: static RAM (SRAM), and dynamic RAM (DRAM).
- ROM memories are classified basically by how they are programmed: masked ROMs, PROMs, EPROMs, EEPROMs, and flash memories.
- Disks are secondary memory devices and can take the form of hard drives and floppy disks.
- Compact disks (CDs) are very similar to magnetic disks except that they use light to store the data on a disk instead of magnetics.
- RAM disks are small removable devices that plug into the USB port and appear to a PC as a "removable drive."

- Magnetic tapes are still used in some instances, mainly for backup purposes.
- There are two general methods by which microprocessor-based systems handle the peripheral I/O devices: peripheral-mapped I/O and memory-mapped I/O.
- Input devices are peripheral devices that get data into the microprocessor system.
- Output devices are peripheral devices that take data from the microprocessor and either display it, store it, or control external processes with it.

Questions

1. What three buses make up the system bus?
2. What are the internal components of a typical microprocessor?
3. What functions does the ALU perform?
4. What component sends out the internal control functions based upon the operation being performed?
5. Name four special-purpose registers in a typical microprocessor.
6. What determines the size of the addressable memory range?
7. Why must the data bus be bidirectional?
8. Why are multiplexed buses used?
9. What are the advantages/disadvantages of SRAM as compared to DRAM?
10. Name four secondary storage memory devices.
11. What is the difference between EPROM and EEPROM?
12. Explain why disks are considered semi-sequential access devices.
13. Explain the difference between peripheral-mapped I/O and memory-mapped I/O.
14. List five different input devices.
15. List three output devices.

3

Programming a General Microprocessor

Objectives:

Upon completion of this chapter, you should:

- Understand the programming model of a typical microprocessor

- Know the classifications of instructions for a typical microprocessor

- Be able to program a generic microprocessor using a generic microprocessor instruction set

- Be able to write, assemble, and execute a generic program

- Know how to flowchart a program

- Understand the use of pseudo-code as a tool

Key Terms:

- **Programming model**—Shows the general-purpose registers, the special-purpose registers, and the size of the buses of a specific microprocessor

- **Data transfer (copy) operations**—instructions that move data between registers, between memory and a register, between a register and memory, between an I/O device and the accumulator, and between the accumulator and an I/O device

- **Arithmetic operations**—Instructions that perform operations such as add, subtract, multiply, divide, increment, and decrement

- **Logical operations**—Instructions that perform operations such as AND, OR, XOR, rotate, compare, and complement

- **Branching operations**—Instructions that change the flow of the program either conditionally or unconditionally

- **Machine control operations**—Instructions that control the actual operation of the microprocessor

- **Flow chart**—A graph which displays the logical flow of a program and assists in the writing of that program

- **Pseudo-code**—The technique of writing the steps of the solution to a problem in regular English phrases without worrying about context, format, or style

Introduction

A microprocessor is programmed using assembly language, and though such programming is specific to the microprocessor being used, all assembly languages are fairly similar and perform the same basic instructions. Some microprocessor assembly languages contain a larger number of instructions than others, but nevertheless a basic set of instructions is common to them all.

Thus, once a person learns to program one microprocessor using its assembly language, learning the assembly languages of other microprocessors is quite easy. Understanding how to write assembly language programs is more about learning to "see" and write programs at a low-level (register and bit-twiddling levels), not at high-level views like with C or Fortran.

In this chapter, we will look at how to program a general microprocessor, not a specific one. The assembly language used here is made-up and is not specific to any one microprocessor. Learning this general assembly language is like using the pseudo-code technique so often employed in high-level language programs. Pseudo-code is a technique that involves using English language instructions when writing a program, which are similar to actual instructions. It allows the person to just write the program in simple English, putting down the intent of the operations on paper without having to use the exact instruction mnemonics and format specific to that particular language. One pseudo-code "instruction" may translate into one actual instruction, or it may expand into more than one.

A later chapter, however, will discuss the exact assembly language used for the 8085 microprocessor.

3-1 The Microprocessor Programming Model

Figure 3-1 shows a **programming model** of a typical microprocessor, which illustrates the general-purpose registers, the special-purpose registers, and the size of the buses. All of these are used by the assembly language programmer.

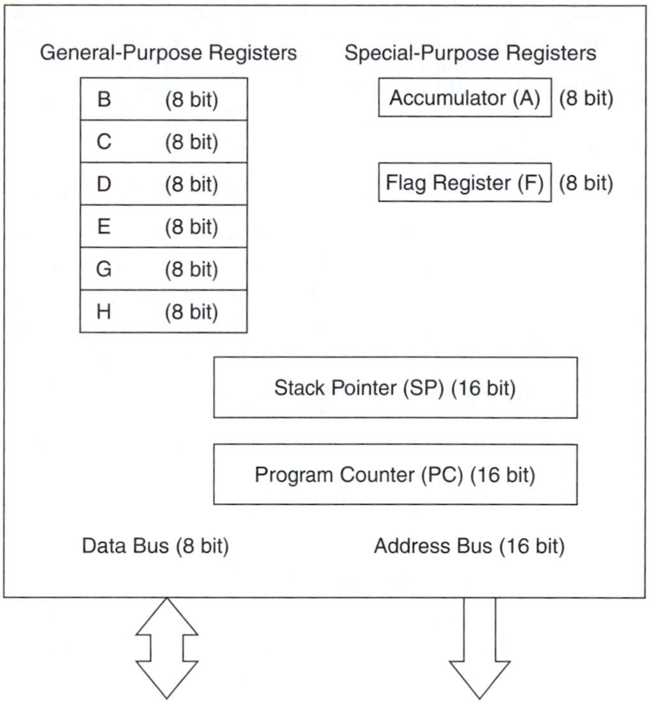

Figure 3-1 ■ A typical microprocessor programming model

In this generic microprocessor model, the general-purpose registers are 8 bits wide, as is the data bus. Two of the special-purpose registers—the accumulator and the flag register—are also both 8 bits wide. Two other special-purpose registers, however—the stack pointer and the program counter—are 16 bits wide in order to match the address bus which is also 16 bits wide. This is a fairly typical microprocessor programming model, but specific microprocessor models should be similar.

The bit level layout of the flag register may vary from one microprocessor to another, but it should contain the same basic information.

3-2 Classifications of Instructions

Assembly language instructions can be classified into five distinct operational categories: **data transfer (copy) operations, arithmetic operations, logical operations, branching operations,** and **machine control operations**. Although the actual instructions vary per microprocessor, they all fall into these general categories.

Figure 3-2 shows the breakdown of instructions into classes.

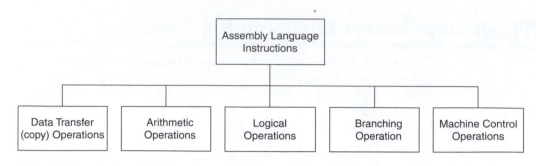

Figure 3-2 ■ Instruction classes

Data Transfer (Copy) Operations

Data transfer instructions move data between registers, between memory and a register, between a register and memory, between an I/O device and the accumulator, and between the accumulator and an I/O device. The accumulator is generally used as the source and or destination for all I/O operations.

Arithmetic Operations

Arithmetic instructions perform operations such as add, subtract, multiply, divide, increment, and decrement. Arithmetic operations usually take place in the ALU, and the result of the operation ends up in the accumulator.

Logical Operations

Logical instructions perform operations such as AND, OR, XOR, rotate, compare, and complement. Logical operations also use the ALU. The results of the logical instructions stay in the accumulator.

Branching Operations

Branching instructions change the flow of the program either conditionally or unconditionally. Jump instructions can either be unconditional or they can be conditional on values as a zero or non-zero result of an operation, a carry flag set, a parity bit set, and so on. These conditions show up as flag bits in the flag register.

Other branching instructions include call and return instructions that are used when going to a subroutine and returning from a subroutine. A subroutine is a single block of code that can be "called" or transferred to from any place in the program, and then when the code in the subroutine is finished, the microprocessor returns to the spot in the program from which the subroutine was called.

Machine Control Operations

Machine control instructions are few in number and control the actual operation of the microprocessor. Examples of machine control operations are halt instructions, interrupt instructions, and no-op instructions. No-op instructions are instructions that do nothing but take up machine cycles. Typically, they're used in delay loops or in other situations that require a fixed number of machine cycles.

3-3 The Instruction Format

The format of a typical instruction is composed of two parts: an operation code or opcode, and an operand. Every instruction needs an opcode to specify what the operation of the instruction is, and then an operand that gives the appropriate data needed for that particular operation code.

Figure 3-3 shows a typical instruction format.

Operation Code	Operand

Figure 3-3 ■ A typical instruction format

Instructions are typically 1 byte, 2 bytes or 3 bytes in length for an 8-bit microprocessor. In an 8-bit microprocessor, such as the 8085, since the word length is 8 bits, the instructions will be 1 byte minimum, up to typically 3 bytes. In the case of 8-bit microprocessors, the operation code or opcode is typically 1 byte or 8 bits.

Instructions can be an opcode only, and hence would only be 1 byte in length, or an instruction can be two or 3 bytes in length where there is an opcode, plus an 8-bit or 16-bit operand. If the operand is data, it is typically 8 bits in length. When the operand is an address, it is 16 bits or 2 bytes in a microprocessor (like in the 8085, which has 16-bit addresses).

1-byte instructions are opcode-only instructions which do not require any other data or address. Examples would be "Complement", "Add a register to the accumulator", or "Move register to register". All of the information needed for these types of instructions is included in the opcode. The registers involved in the operation are encoded into the opcode.

2-byte instructions contain an opcode plus 8 bits of data. Examples would be "Move data into a register", "Add data to a register", or "Subtract data from a register".

3-byte instructions contain an opcode plus a 16-bit address. Examples would be "Load contents of a memory location at a given address into a register", "Store contents of a register at an address", and "Jump to an address".

For different microprocessors, the length of instructions will vary to match that particular microprocessor. Usually the opcode is the same length as the width of the data registers or data bus in the microprocessor, data operands will be the same width as the data bus, and address operands will be the same length as the width of the address bus.

3-4 A Generic Microprocessor Instruction Set

Let's now discuss a typical instruction set for a generic microprocessor. Note, however, that any instruction set for a specific microprocessor will be similar to this generic instruction set.

Data Transfer (Copy) Instructions:

All have the form:

Opcode destination, source

Examples of data transfer instructions:
- MOVE Reg, Reg
- MOVE Reg, Data
- LOAD Reg, Memory
- STORE Memory, Reg
- IN Reg, I/O Port
- OUT I/O Port, Reg

Arithmetic instructions:
- ADD Reg (to Acc)
- SUBTRACT Reg (from Acc)
- INCREMENT Reg (by 1)
- DECREMENT Reg (by 1)

Logical instructions:
- AND Reg (with Acc)
- OR Reg (with Acc)
- XOR Reg (with Acc)
- COMPARE Reg (with Acc)

Branch instructions:
- JUMP Address
- JUMPZ Address (if Acc is zero)
- JUMPNOTZ Address (Acc is not zero)
- JUMPC Address (if carry flag is set)
- JUMPNOTC Address (if carry flag not set)
- CALL Subroutine Address
- RETURN (return from called subroutine)

Machine Control Instructions:
- HALT (stop processor)
- NOOP (no operation)

Most microprocessor instruction sets have more instructions in them than this, and the instructions are much more complex than these. However, all microprocessor instruction sets will have these basic operations in them in some form or another.

Also the format of "opcode destination, source" may be switched to "opcode source, destination." The Intel microprocessors typically use "opcode destination, source", while the Motorola microprocessors typically use "opcode source, destination."

3-5 Writing, Assembling, and Executing a Program

Figure 3-4 shows the general flow of the process by which we write, assemble, and then execute a microprocessor program.

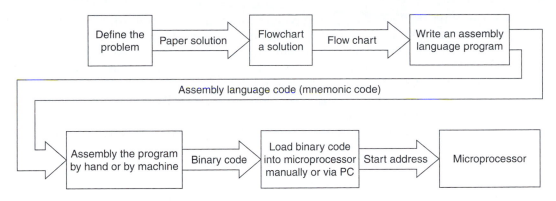

Figure 3-4 ■ Writing, assembling, and executing a program

Define the Problem

In creating a program to run on a microprocessor, the first step is to define the problem. You can do this by listing the inputs, noting the outputs expected, and finally the operation(s) to be performed. Once you've written down the problem in that form (inputs, outputs, and processes to be performed), you can begin to design a solution.

Flowchart a Solution

In designing a solution for the problem, the **flow chart** is a very useful tool/technique. How to flowchart a program will be discussed in more detail in the next section, but basically a flow chart allows you to begin designing the solution at a very high level. The easiest is to start with a very simple, high-level flow chart. At a high level, each box in a flow chart can do anything—getting inputs, doing an operation, or putting out results. It should just show the sequence of events that need to happen to solve the problem. At this high level, the boxes should not be related to specific assembly language instructions. That is a lower level and comes later.

After creating a high-level flow chart, you can then break down each box in the flow chart into smaller steps. This process can be repeated until each box corresponds to a single assembly language command.

Write an Assembly Language Program

Once you have a low-level flow chart, you can convert it to an assembly language program. Each box in the flow chart can be represented by one or more assembly language instructions. The actual mnemonics and commands that you use depend upon which specific microprocessor you're writing the program for. Nevertheless, the process is the same. As you're writing the assembly language program, you'll need to decide which registers should be used for each operation, as well as which memory locations. Then you can write the program in the commands specific to that microprocessor. Attention needs to be paid also to the exact syntax for the commands themselves.

Assemble the Program

The program can be assembled by hand, if it is small enough, or by a cross-assembler if it is a larger program. To assemble a program by hand, you look up the hexadecimal equivalent for each instruction in your program. The hex form is easier to work with than binary. Although you are converting the assembly language program into a binary program, you represent the binary equivalent of each instruction in hex. The data operands and the address operands are also converted to hex.

If you use a cross-assembler, you basically run a program—on a PC, for example—which takes the assembly language program as input, and then automatically generates the binary form of the program.

Load and Execute the Program

If you hand-assembled the program, you then load the program into the microprocessor from a keypad in hex form, byte by byte. You begin loading the program at a specific start address. This start address is determined by the type of microprocessor you're using and what the application is—for example, whether it is a trainer with a utility program, or a production microprocessor with an operating system, or a bare machine with no other program. You then load the start address of your program into the program counter, and start the microprocessor. If it is a bare machine instead, you load your program so that when the microprocessor is reset, it goes to the start of your program.

If you used a cross-assembler, you load the program over a serial link between the PC and the microprocessor, specifying in the cross-assembler where the program is to be loaded—that is, the origin of the program. The cross-assembler then sends the binary program to the microprocessor, which loads it starting at the specified origin. Executing the program is the same as in the hand-assembled case.

3-6 Flowcharting a Program

Flowcharting is a technique or tool to use when writing programs of any kind. It can be used for assembly language programs or high-level programs of any language. It's basically a method that's used to design a programming solution to solve a specific problem.

Only a few basic blocks are used in a simple flow chart: an operation block, an input/output block, a subroutine block, a decision diamond, and a start and end oval.

Figure 3-5 shows the basic building blocks used in flow charts.

These basic blocks can be used to flowchart any solution to a programming problem.

The Operation Block

The operation block can be used for any operation, arithmetic or logical calculation, or function. It is the general purpose "block."

When you do a high-level flow chart, you may only use operation blocks, and then in successive iterations break the high-level blocks into more specific flow-chart blocks.

The Subroutine Call

The subroutine call block is just that—a subroutine call. The name of the subroutine called is usually put in the upper part of the box, and often (but not always) the input parameters for the subroutine are listed in the box.

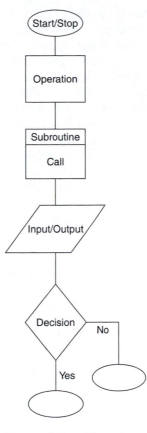

Figure 3-5 ■ Flow chart building blocks

Subroutines are blocks of code that exist in a program which are transferred to from any point in the program. When the subroutine is finished, the microprocessor returns to the part of the program where the transfer to the subroutine occurred. The purpose of a subroutine is simply to reduce the amount of code needed by allowing a function that is used repeatedly by different parts of the program to be written once, and then be transferred back and forth whenever needed.

Input/Output Block

The input/output block indicates an operation of either reading in data from some input, or putting out data via some output device. Typically, when the peripheral-mapped I/O method is used, these blocks indicate an input or output instruction.

Decision Block

The decision block is used whenever a question needs to be asked or a decision must be made on some piece of data. For example—Go to A if the accumulator is zero, but go to B if the accumulator is not zero.

Start/Stop

The start/stop symbol is nothing other than a block or symbol that indicates the beginning or end point of a routine or program. If the program is large enough to span more than one page, a connector symbol (basically a start/stop symbol) can be used to show where the program goes to on the next sheet.

When flowcharting a programming solution to a problem, whether in assembly language or a higher-level language, you start with a high-level solution using as few blocks as needed, where the operations within each block are high level and don't sink to the instruction level. Then, in successive iterations, you keep breaking up each block in the flow chart into lower-level operations until you get to a low level flow chart where each block is basically one assembly language instruction.

Example 3-1 shows a problem that can be first solved with a high-level flow chart, and then shows a detailed, low-level flow chart arrived at by breaking down the operations in the high-level blocks into assembly language level. Then, the assembly language program is written in the generic language defined earlier in this chapter.

Example 3.1

Problem: Write an assembly language program that inputs two numbers from memory (locations 2050H and 2051H), calculates the sum of the two numbers and stores it in memory (at 2052H), and calculates the difference between the two numbers and then stores that in memory (at 2053H) as well.

Solution: A High-Level Flow Chart (see Figure 3-6):

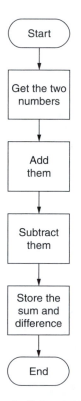

Figure 3-6 ■ Example 3.1—A high-level flow chart

A Detailed Flow Chart (see Figure 3-7):

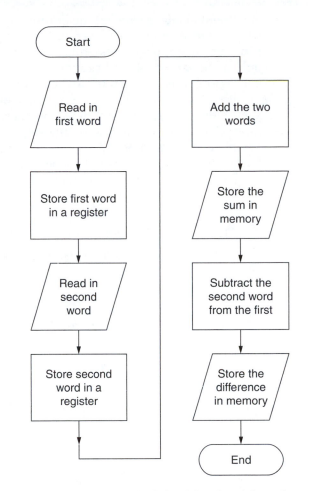

Figure 3-7 ■ Example 3.1—A low-level flow chart

An Assembly Language Program:

    LOAD A,2050H (load first word located in memory location 2050Hex into accumulator)

    MOVE B,A (copy first word into register B)

    LOAD A,2051H (load second word into the accumulator)

    MOVE C,A (copy second word into register C)

    MOVE A,B (copy first word into the accumulator)

    ADD C (add two words)

    STORE 2052H,A (store sum into memory location 2052H)

    MOVE A,B (copy first word into the accumulator)

    SUBTRACT C (subtract second word from first word)

    STORE 2053H,A (Store difference at location 2053H)

3-7 Assembly Language Pseudo-Code

Pseudo-code is a technique or tool for facilitating the writing of a program. It's most often employed when writing programs in higher-level languages and is not usually applied to assembly language programs. However, pseudo-code is only a software tool, and can just as easily be used for writing assembly language programs as it can for higher-level programs.

Pseudo-code is simply the technique of writing the steps of the solution to a problem using regular English phrases without worrying about context, format, or style. You just write the steps that are needed to solve the problem. Pseudo-code is commonly used in place of a high-level flow chart, which can be developed from the pseudo-code later.

Figure 3-8 shows the pseudo-code solution to Example 3.1.

```
- Get the two numbers from memory

- Add the two numbers together

- Store the sum in memory

- Subtract the two numbers

- Store the difference in memory
```

Figure 3-8 ■ Pseudo-code for Example 3.1

This is a fairly simple example, and the pseudo-code seems obvious. But, in a more complex problem the use of pseudo-code instead of a high-level flow chart may give the designer an easier way to develop the high-level steps needed to solve the problem.

Pseudo-code is presented as a technique often used in high-level language program development that can be employed in lieu of high-level flowcharting, or it can be used first, followed by a medium-level flow chart and then a detailed flow chart.

Example 3.2 shows pseudo-code being used to develop the steps for the solution of the problem.

Example 3.2

▼

Problem: If an input switch is set, read in the value of a temperature sensor, convert the value from binary to BCD, and then display the value on LEDs.

Solution: Pseudo-code:

    Check to see if the input switch is set
    If it isn't set, do nothing and check the switch again
    If the switch is set, read in the binary value of the sensor
    Convert the binary value to BCD
    Display the BCD value on the LEDs

▲

Example 3.2 could also be solved using a high-level flow chart.

Summary

- Assembly language is used to program microprocessors.
- A microprocessor programming model is one that shows the general-purpose registers, the special-purpose registers, and the size of the buses.
- Assembly language instructions can be classified into five distinct operational categories: data transfer (copy) operations, arithmetic operations, logical operations, branching operations, and machine control operations.
- Data transfer instructions move data between registers, between memory and a register, between a register and memory, between an I/O device and the accumulator, and between the accumulator and an I/O device.
- Arithmetic instructions perform operations such as add, subtract, multiply, divide, increment, and decrement. Arithmetic operations take place in the ALU.
- Logical instructions perform operations such as AND, OR, XOR, rotate, compare, and complement.
- Branching instructions change the flow of the program either conditionally or unconditionally.
- Machine control instructions are few in number and control the actual operation of the microprocessor.
- The format of a typical instruction is two parts: an operation code or opcode and an operand.
- Instructions are typically one byte, 2 bytes, or 3 bytes in length for an 8-bit microprocessor.
- In creating a program to run on a microprocessor, the first step is to define the problem by listing the inputs, listing the outputs that are expected, and then outlining the operation(s) that are to be performed.
- In designing a solution for a problem, the flow chart is a very useful tool or technique.
- An assembly language program can be assembled by hand if it is small, or by a cross-assembler if it is a larger program.
- Only a few basic blocks are used in a rudimentary flow chart: a simple operation block, an input/output block, a subroutine block, a decision diamond, and a start and end oval.
- Pseudo-code is simply a technique of writing out the steps for solving a problem using regular English phrases without any worry about context, format, or style. It is useful for writing a program.

Questions

1. In a typical microprocessor model, what determines the width of the general-purpose registers?
2. In a typical microprocessor model, why are the stack pointer and the program counter the same width as the address bus?

3. Why is the accumulator so special?

4. Name the five categories that assembly language instructions can be classified into.

5. What category does the HALT instruction fall into?

6. Give an example of a data transfer instruction.

7. What are the two basic ways a branching instruction can operate?

8. What are the two basic parts of a typical instruction?

9. How many bytes in length is a typical instruction in an 8-bit microprocessor that has an address as the operand?

10. What type of instruction is usually 1 byte in an 8-bit microprocessor?

11. Name a major difference in how Intel and Motorola format their instructions.

12. What is a cross-assembler?

13. What does hand-assembling mean?

14. Name and draw the five major building blocks used in flowcharting.

15. What is pseudo-code?

Problems

1. Draw a flow chart for a program that takes two numbers stored in memory at locations 3050H and 3051H, adds them together, and then stores the sum at memory location 3052H.

2. Write the assembly language program (using the generic assembly language instruction set in this chapter) to perform the operation defined in Problem 1.

3. Write an assembly language program to compare two numbers stored in memory at locations 303AH and 303BH, which—if they are equal—transfers to a program at address 2020H, and if they are not equal, transfers to a program at address 2050H. First draw a flow chart for this program.

4. Write the pseudo-code for a program that reads in a number from memory location 2050H, decrements it by one, and when that number reaches zero, sends FFH out on I/O Port 01.

5. Write the assembly language program for the problem defined in Problem 4.

Laboratory Experiments

Assumptions to Make with Regards to the Following Labs:

- The labs are to be run on an 8085 microprocessor trainer.
- The trainer should have keyboard entry and LED display.

- The trainer has 2000H as the starting address for user RAM. If your trainer has a different starting address for users, just substitute that base address for 2000H.
- Users should be able to input hexadecimal values to enter a program.
- Hand-assembling is assumed, but if your trainer is connected to a PC and a cross-assembler is available, make the appropriate wording changes in the labs to accommodate this.
- The student can view the contents of memory and registers on the LED display on the trainer. If, however, the student is using a PC interface, make appropriate changes to recognize that the students will be viewing the memory and register contents on the PC.
- The trainer is assumed to have a single-step capability.
- Lab 10 assumes the availability of an 8255 PPI chip and the ability to connect it up to the 8085 trainer.
- Lab 11 assumes a basic logic analyzer is available, such as the Tektronix 318/338 Logic Analyzer.
- Lab 12 assumes the availability of a basic PC-based 8085 microprocessor simulator.

The use of a microprocessor programming sheet and a tracing sheet are assumed. These forms are used when writing a program and when tracing register and memory contents. A sample programming sheet and tracing sheet are attached at the end of this section.

This set of labs is intended to cover a complete 16-week semester where there is one three-hour lab session per week. There are 12 labs, and Labs 9 and 10 are two-week labs, so that covers 14 weeks. Usually, the first lab session is not used, and the last lab session is used for make up.

Lab 1: Introduction to the Trainer

Introduction: This lab will introduce the student to the microprocessor trainer.
Objectives: Upon completion of this lab, the student will:

- Be familiar with how to power on and reset the microprocessor trainer
- Be able to enter data into the RAM memory
- Be able to read the registers in the microprocessor
- Be able to single-step the trainer

Equipment Needed:

- An 8085 microprocessor trainer

Procedure:

1. Connect the power supply to the microprocessor trainer and apply power.
2. Push the Reset button on the trainer so that it resets and comes up available for use.
3. After reading the trainer's manual about how to enter data into memory, load the value 05H into memory location 2000H. Then load A0H into memory location 2001H.

4. Consult the trainer's manual on how to view the contents of registers and memory and afterward view the contents of memory locations 2000H and 2001H to verify that 05H and A0H are still there. Record these values.

5. Load the following instructions into memory, starting at location 2000H:

Machine Language	Assembly Language	Comment
3E FC	MVI A, FCH	; move FCH into register A
16 01	MVI D, 01H	; move 01H into register D
76	HLT	; halt

   When entering the preceding code, input the hexadecimal values of the machine language and start loading them, a byte at a time, at location 2000H.

6. Single-step the trainer, starting at location 2000H, until the HLT is executed.

7. View the contents of registers A and D on the LED display and record their values.

8. Load the following instructions into memory, starting at location 2000H:

Machine Language	Assembly Language	Comment
3E 01	MVI A, 01H	; load 01H into A
17	RAL	; rotate A left one place
17	RAL	; rotate A left one place
3E FF	MVI A, FFH	; load FFH into A
76	HLT	; halt

9. Single-step the trainer, starting at location 2000H. After each step, observe the contents of A and record it. Continue until the halt is reached.

Questions

1. What ends up in the accumulator after the following code is executed?

   MVI A, 09H
   RAR
   HLT

2. How many bytes of memory are needed to store the code listed in Question 1?

3. How do you view the contents of a specific register in the trainer?

4. What sequence of steps do you need to go through in order to load a program and then single-step the trainer?

5. Write a brief conclusion for this lab.

Lab 2: Data Transfer (8-Bit) Instruction Lab

Introduction: This lab will demonstrate the transfer of data into and out of the 8-bit registers.

Objectives: Upon completion of this lab, students will:

- Be able to hand-assemble a program
- Be able to identify the following addressing modes: immediate, register, direct, and indirect
- Be familiar with the 8-bit data transfer instructions
- Be able to predict and monitor changes in the contents of registers and memory locations as instructions are executed

Equipment Needed:

- An 8085 microprocessor trainer

Procedure:

1. Hand-assemble the following program, putting it on a microprocessor programming sheet. Show the address, machine language, assembly language, and comments. Start the program at location 2000H.

   ```
   START:  LDA 2020H      ; load 2020H into A
           MVI B, 5AH     ; load B with 5AH
           MOV A, B       ; move B into A
           STA  2021H     ; store A at location 2021H
           HLT            ; halt
   ```

2. Load memory location 2020H with 4FH. Clear memory location 2021H. Make sure registers A and B are clear.

3. Predict which registers and memory locations will change after each instruction. Execute one instruction by single-stepping the trainer. Examine the registers that have changed, as well as any memory locations that have changed, and record these values in the tracing worksheet.

4. Repeat step 3 for each step in the program until the HLT is reached. Note: The program counter will always be pointing to the next instruction to be executed, so when you record the value of the program counter on the tracing worksheet, the values of the registers and memory locations recorded will be the result of the previous instruction—not the instruction the program counter will be pointing to.

5. Hand-assemble the following program and record it on a microprocessor programming sheet. Load the program into the trainer starting at location 2000H.

   ```
   START:  LXI  H, 2020H    ; load memory address 2020H into HL
           MOV  A, M        ; move contents of location 2020H into A
           MVI  B, AAH      ; load B with AAH
           LXI  H, 2021H    ; load memory address 2021H into HL
           MOV  M, B        ; load location 2021H with contents of B
           HLT              ; halt
   ```

6. Load memory location 2020H with the value 77H.

7. Step through the preceding program, one instruction at a time. Record the values of the registers and memory locations that change on each instruction. Use the tracing worksheet.

Questions

1. List an example of direct addressing in one of these two programs.
2. List one example of register addressing.
3. List one example of immediate addressing.
4. List one example of register indirect addressing.
5. Write a brief conclusion for this lab.

4

The Hardware Architecture
of the 8085 Microprocessor

Objectives:

Upon completion of this chapter, you should:

- Understand the hardware architecture of the 8085

- Comprehend the internal architecture and bus sizes of the 8085

- Be familiar with the interrupt capabilities of the 8085

- Understand bus timings and how instructions break down into machine cycles

- Be able to interface memory to the 8085

- Understand memory address decoding

- Know how to interface I/O to the 8085

Key Terms:

- **8085 microprocessor**—This is an improved version of its predecessor, the 8080A microprocessor. The 8085 is actually the 8085A, but is commonly referred to as the 8085.

- **Multiplexed bus**—A bus where the data bus is overlapped or shared with the lower part of the address bus

- **SID (serial input data)**—An input port on the 8085 which is used for inputting serial data and is accessed by special instructions

- **SOD (serial output data)**—An output port on the 8085 which is used for outputting serial data and is accessed by special instructions

- **Arithmetic logic unit (ALU)**—Performs the arithmetic and logical operations

- **Stack pointer**—A register that points to the next available slot in the stack

- **Program counter**—A register that holds the address of the next instruction to be executed

- **Instruction register**—Part of the instruction decoding circuitry; it's where an instruction is read into from memory.

- **Interrupt control unit**—Handles all of the incoming interrupts to the microprocessor, as well as the outgoing interrupt acknowledgement signal

- **Serial I/O control unit**—Handles the two serial I/O pins SID and SOD, which allow serial data into, as well as out of, the microprocessor

- **Opcode fetch**—The first machine cycle in every instruction; it is where the opcode is fetched from memory and returned on the data bus to the microprocessor.

- **Memory read cycle**—When the microprocessor reads in data from memory

- **Memory write cycle**—When the microprocessor sends data out from the accumulator or a specific register, and writes it into memory

- **I/O read cycle**—When an IN instruction is executed by the microprocessor, and data is read in from an I/O device

- **I/O write cycle**—When an OUT instruction is executed by the microprocessor and data from the accumulator is written out to the I/O device specified by the port address

- **Interrupt acknowledge cycle**—A special machine cycle that is used in place of the opcode fetch cycle in the RST (restart) instruction

- **Memory map**—A diagram that shows all of the possible addresses in a microprocessor system and what they are assigned to

- **Memory-mapped I/O**—The method of addressing I/O devices that includes the addresses of the I/O devices within the memory address range and therefore treats them like any other memory location

- **Peripheral-mapped I/O**—The method where the address range for the I/O devices is outside of the address range for the memory of the system, and only I/O instructions can be used

Introduction

The **8085 microprocessor** by Intel is an improved version of its predecessor, the 8080A. It's actually the 8085A, but is commonly referred to as the 8085. The 8085 contains most of the logic circuitry for performing computing tasks and communicating with peripherals on its single chip.

The 8085 is an 8-bit microprocessor. The data bus is 8 bits wide, while the address bus is 16 bits wide. The internal general-purpose registers are also 8 bits wide to match the width of the data bus. Because of its 16-bit address bus, however, the 8085 can address 64K bytes of memory.

The 8085 has a **multiplexed bus**, which means that the 8-bit data bus is overlapped or shared with the lower 8 bits of the 16-bit address bus. This feature is usually used when the chip designers run out of I/O pins on the chip and so are forced to multiplex or share pins between the data bus and part of the address bus. It would be simpler if the data bus and the address bus were totally separate buses; however, when multiplexed, they are not. This means that when the microprocessor uses the address bus, it can't utilize the data bus, and vice versa. Thus, the microprocessor operation must take this multiplexing into account when using the buses.

4-1 The 8085 Microprocessor

The 8085 is an 8-bit microprocessor with an 8-bit data bus and a 16-bit address bus, allowing it to access 64K bytes of memory (2^{16} = 65,536 or 64K).

The chip itself has 40-pins and can work with a clock up to 3MHz.

Figure 4-1 shows the 8085 chip layout, with the pins labeled. All of the signals can be put into six groups: (1) address bus, (2) data bus, (3) control and

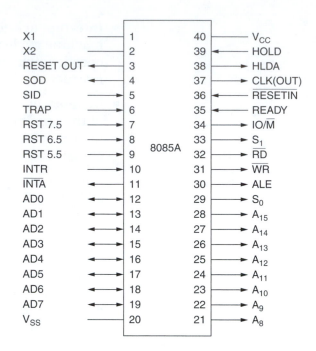

Figure 4-1 ■ The 8085 microprocessor pinout

status signals, (4) power supply and clock signals, (5) externally initiated signals, and (6) serial I/O ports.

Address Bus

The 8085's address bus is 16 bits wide and can address 64K bytes of memory. It is split into two groups: AD0–AD7 and A8–A15. The low-order 8 bits of the address bus are labeled AD0–AD7 because they are multiplexed with the 8-bit data bus, while the upper 8 bits of the address bus are labeled A8–A15.

The high-order section of the address bus is unidirectional out, while the lower 8 bits AD0–AD7 are unidirectional when used as part of the address bus.

Data Bus

The data bus is 8 bits wide and is labeled AD0–AD7 since the data bus is shared or multiplexed with the low order bits of the address bus. When used as the data bus, AD0–AD7 is bidirectional.

Control and Status Signals

This group of signals includes two control signals ($\overline{RD}$ and $\overline{WR}$), three status signals (IO/$\overline{M}$, S_1, and S_0), and one special symbol (ALE).

- $\overline{RD}$ **(Read)** This is the read control signal: active low. When it goes low, it indicates that the selected memory or I/O device is to be read and that the data will be coming in on the data bus.

- $\overline{WR}$ **(Write)** This is the write control signal: active low. When it goes low, it indicates that the selected memory or I/O device is to be written and that the data to be written is on the data bus.

- **IO/$\overline{\text{M}}$** This status signal indicates whether the operation being performed is an I/O operation (IO/$\overline{\text{M}}$ line high) or a memory operation (IO/$\overline{\text{M}}$ line low).
- **S_1 and S_0** These status signals further indicate the state of the microprocessor, but they are rarely used in small and medium systems. (Table 4.1 shows the states of S_1 and S_0 for different machine cycles.)
- **ALE (address latch enable)** This is a positive going signal generated every time the 8085 starts an operation (machine cycle). It indicates that AD0–AD7 are address bits. This signal is often used to latch the low-order address bits from the multiplexed bus.

Power Supply and Clock Signals

The power supply and clock signals are listed next:

- **V_{CC}** The +5V power supply
- **V_{SS}** The ground reference
- **X_1 and X_2** A crystal or clock circuit is connected at these pins. The frequency used internally is half of the frequency input, so to run at 3MHz, the crystal or clock circuit should have a frequency of 6MHz.
- **CLK (OUT): (clock output)** This is an output of the 8085's clock and can be used as a clock for external devices.

Externally Initiated Signals

The 8085 has five interrupt signals that can be used by external devices to interrupt a program that is executing. They are INTR, RST7.5, RST6.5, RST5.5, and TRAP. The INTA is the interrupt acknowledgement signal used by the 8085.

- **INTR (interrupt request)** This is used as a general-purpose interrupt.
- **RST7.5, RST6.5, RST5.5 (restart interrupts)** These are vectored interrupts that transfer the program control to specific memory locations called Interrupt vector locations. They are higher priority than the INTR interrupt, and 7.5 is higher than 6.5 which is higher than 5.5.
- **TRAP** A nonmaskable interrupt that is the highest priority of all interrupts. Nonmaskable means it cannot be blocked by a program like other interrupts.
- **$\overline{\text{INTA}}$ (interrupt acknowledgement)** An output signal that is used to acknowledge an interrupt.

The other pins in this group are $\overline{\text{RESETIN}}$, RESET OUT, READY, HOLD, and HLDA. They are inputs and outputs for use by external devices.

- **$\overline{\text{RESETIN}}$ (Reset)** When this input goes low, the program counter is set to zero, the buses are tri-stated, and the microprocessor is reset.
- **RESET OUT** This is an output signal indicating that the microprocessor is being reset. This signal can be used to reset other devices.

- **READY** This input signal is used to delay the microprocessor Read or Write cycle in the cases where a slower peripheral device is not ready for the read or write to take place. When READY goes low, the microprocessor waits until READY goes high.
- **HOLD** This input signal indicates that a peripheral device such as a DMA controller wants to use the data and/or address buses.
- **HLDA (hold acknowledge)** This signal is used to acknowledge the receipt of a hold request.

Serial I/O Ports

SID (serial input data) and **SOD (serial output data)** are the I/O ports on the 8085 that are used for inputting and outputting serial data. They are accessed via specific instructions in the 8085 instruction set to send out and receive serial data.

This accounts for all 40 pins on the 8085.

Table 4.1 lists the states of the status signals and control signals for the different machine cycles.

	Status	Signals		Control	Signals	
Machine Cycle	IO/$\overline{\text{M}}$	S_1	S_0	$\overline{\text{RD}}$	$\overline{\text{WR}}$	$\overline{\text{INTA}}$
Opcode Fetch	0	1	1	0	1	1
Memory Read	0	1	0	0	1	1
Memory Write	0	0	1	1	0	1
I/O Read	1	1	0	0	1	1
I/O Write	1	0	1	1	0	1
Interrupt Ack	1	1	1	1	1	0
Halt	Z	0	0	Z	Z	1
Hold	Z	X	X	Z	Z	1
Reset	Z	X	X	Z	Z	1

Notes: Z = tri-state (high impedance)
X = unspecified

Table 4.1 ■ The 8085 Machine Cycle Status and Control Signals

The 8085 Internal Hardware Architecture

Figure 4-2 is a functional block diagram for the internal architecture of the 8085 microprocessor.

The basic blocks in the 8085's internal architecture are the arithmetic logic unit (ALU), timing and control, register array, instruction decoder, interrupt control, and the serial I/O control.

The Arithmetic Logic Unit (ALU)

The **arithmetic logic unit (ALU)** performs both the arithmetic and logical operations. The accumulator is also part of the ALU, as is the temporary register and five flags. The results of most operations are stored in the accumulator, while the five flags reflect the status of the resulting data in the accumulator.

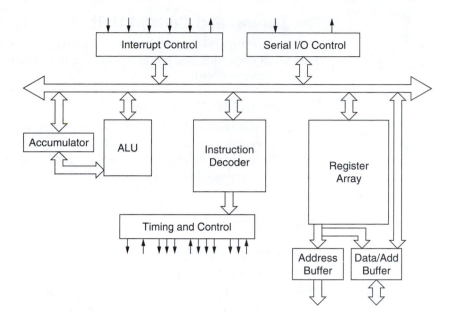

Figure 4-2 ■ An 8085 functional block diagram

The five flags are

- **S (sign flag)** After the end of a logical or arithmetic operation, if the highest order bit D7 in the accumulator is set, the sign bit is set.
- **Z (zero flag)** If the result of an operation in the ALU is zero, the zero flag is set.
- **AC (auxiliary carry)** This flag is set when a carry is generated between D3 and D4 in the accumulator during the operation. It is only used internally for BCD conversion and is not available for general use by programs.
- **P (parity flag)** This flag is set after an operation when there are an even number of 1's in the result. If there are an odd number of 1's in the result, the parity flag is reset.
- **CY (carry flag)** If the operation performed results in a carry being generated from D7, then this flag is set. Otherwise, the CY flag is reset.

Figure 4-3 shows how these flags are positioned in the flag register.

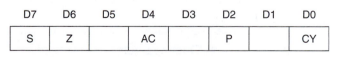

D7	D6	D5	D4	D3	D2	D1	D0
S	Z		AC		P		CY

Figure 4-3 ■ The flag register

The S, Z, P, and the CY flags are all accessible via programs and are used often. The AC flag is only used internally by BCD operations and as such is not available for programming use.

The Timing and Control Unit

The timing and control unit synchronizes all of the microprocessor operations with the clock and generates all the necessary control signals for the rest of the microprocessor.

The Register Array

The register array contains all of the programmable registers B, C, D, E, H, and L, as well as two temporary registers—W and X—that are not accessible to general programming use. The programmable registers used will be discussed more in the next chapter, "Programming the 8085-Basic Commands". The register array also contains the 16-bit stack pointer as well as the 16-bit program counter. The **stack pointer** points to the next available slot in the stack, while the **program counter** holds the address of the next instruction to be executed.

The Instruction Decoder

The **instruction register** is part of the instruction decoding circuitry. When an instruction is read in from memory, it goes into the instruction register. The instruction decoder then decodes the instruction and determines the sequence of operations in the microprocessor that must take place. The instruction register is internal and as such is not addressable via programs.

The Interrupt Control Unit

The **interrupt control unit** handles all of the incoming interrupts to the microprocessor as well as the outgoing interrupt acknowledgement signal.

The Serial I/O Control Unit

The **serial I/O control unit** handles the two serial I/O pins SID and SOD, which allow serial data both into and out of the microprocessor.

Bus Timings and 8085 Machine Cycles

The 8085 microprocessor is designed to execute 74 different instructions. Each instruction is divided into two parts: the operation code (opcode) and the operand. The opcode tells what the operation is (such as ADD or MOVE), and the operand is the necessary information needed for the instruction. It may be data, or an address, or other information needed for the instruction.

Each instruction is broken down into machine cycles, and each machine cycle breaks down into clock cycles. These clock cycles are also called T-states.

The first machine cycle in every instruction is the **opcode fetch**. This is where the opcode is fetched from memory and returned on the data bus to the microprocessor. The subsequent machine cycles depend upon what the fetched opcode is.

Machine cycles break down into the following types:

- Opcode fetch
- Memory read
- Memory write
- I/O read
- I/O write
- Interrupt acknowledge

The Opcode Fetch Machine Cycle

The first operation in every instruction is the *opcode fetch*. This machine cycle retrieves the opcode of the instruction that is being executed. It sends the address out on the address bus of the instruction's opcode and then puts the data (opcode) back onto the data bus.

The opcode fetch cycle is called the M_1 machine cycle and is usually composed of four clock cycles or T-states: T_1–T_4. (A few special 8085 instructions actually have six T-states in their opcode fetch machine cycle.) During T_1–T_3, the address is put out on the address bus and the opcode is returned on the data bus. T_4 is used to decode and execute the opcode. The machine cycles that follow depend upon what the instruction actually is.

Different instructions have different numbers and types of machine cycles in them; the number of T-states per machine cycle also varies. So, every instruction has a specific number and type of machine cycle associated with it, as well as a specific number of T-states.

For example, the STA addr instruction has four machine cycles in it, an opcode fetch cycle, followed by 2 read cycles, and then a write cycle. The total number of T states for all four machine cycles is 13. The instruction set listed in Appendix I shows, along with other information, the number of machine cycles and T-states for every instruction.

To determine which machine cycles are included in an instruction, you only have to determine what operations the instruction must carry out. The first machine cycle is always an opcode fetch. In the case of the STA addr instruction, the second machine cycle needs to be a read cycle to get the first part of the memory address. Then another read cycle is needed to get the second part of the address, and then finally a write cycle should be used to store the contents of the accumulator into the memory address read in. Remember, the data bus is only 8 bits wide, so to read in a 16-bit address takes two memory read cycles.

Example 4.1

▼

Problem: What are the machine cycles and how many T-states are there in the instruction ADI data?

Solution: An ADI data will need an opcode fetch cycle, and then a read cycle to get the data to add in from memory.

For the number of T-states, see Appendix I.

ADI data (add immediate) is a two-byte instruction, (one byte for the opcode and one byte for the data), composed of two cycles and seven T-states. The opcode fetch cycle is four T-states, and the following read cycle is three T-states.

▲

Figure 4-4 shows the timing for the execution of the MVI B,42H instruction, displays the machine cycles and associated T-states for this instruction.

As seen in Figure 4-4, during the first three T-states of the opcode fetch cycle T_1–T_3, the address of the instruction to be fetched (2000H) is put out on the address bus, and the opcode (06H) is returned over the data bus. The fourth T-state (T_4) of the opcode fetch machine cycle is used by the microprocessor to decode and execute the opcode retrieved.

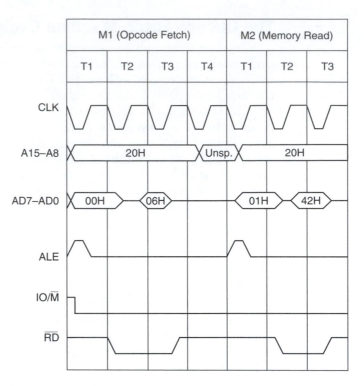

Figure 4-4 ■ The opcode machine cycle timing diagram

The second machine cycle shown in Figure 4-4 is a read cycle that is needed in order to get the data moved into the B register. The address going out over the address bus (2001H) is where the second byte of the instruction is stored, and the data (42H) is returned on the data bus.

Note that the ALE (address latch enable) signal goes high during the beginning of every machine cycle. Also, IO/$\overline{\text{M}}$ goes low at the beginning of the opcode fetch cycle (indicating a memory operation) and stays low during the second machine cycle (a memory read). Also, $\overline{\text{RD}}$ goes low each time there is data on the data bus to be read into the microprocessor—during T_2–T_3 in the opcode fetch cycle, and during T_2–T_3 in the memory read cycle.

Memory Read

A **memory read** machine cycle is a machine cycle during which memory is read. For example, the instruction LDA 2020H, has three memory read cycles following the opcode fetch cycle. The first two memory read cycles are needed to get the memory address, in two 8-bit groups (the low-order part of the address and then the high-order part of the address). The third read cycle is needed to read in the data located at the address previously retrieved. This data is then loaded into the accumulator.

Figure 4-5 shows the timing diagram for an LDA 2020H instruction. It also shows the opcode fetch cycle, followed by the three read cycles. As shown in the figure, after the opcode fetch cycle, the first two read cycles have the address going out over the address bus first for the low-order part of the address (2001H), and then for the high-order part of the address (2002H). In the third read cycle, the address of the instruction just read in from memory

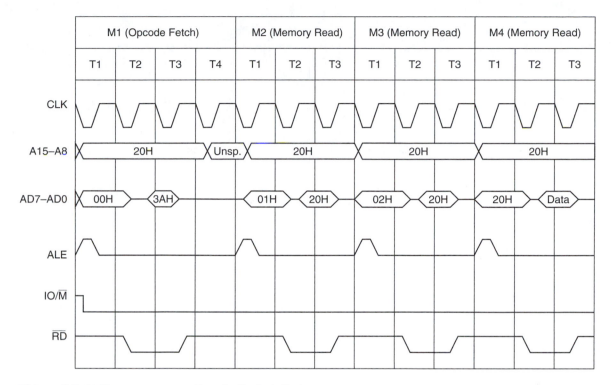

Figure 4-5 ■ The memory read cycle timing diagram

(2020H) is sent back out over the address bus in T_1, and then the data from that memory location is returned over the data bus in T_2–T_3.

$IO/\overline{M}$ goes low at the beginning of the opcode fetch cycle, and remains low during the next three cycles—all of which are memory operations.

$\overline{RD}$, on the other hand, goes low each time data on the data bus is to be read into the microprocessor.

Memory Write

The **memory write** cycle is used when the microprocessor needs to send data out from the accumulator or a specific register, and then write it into memory.

For example, the instruction MOV M,A moves data in the accumulator to the memory location specified by the HL register. This instruction requires two machine cycles—an opcode fetch cycle followed by a write cycle. This is because after fetching the opcode, the instruction has to write the data in the accumulator out to memory at the address located in the HL register. The operation requires seven T-states: four T-states in the opcode fetch cycle and three T-states in the write cycle.

Figure 4-6 shows the timing of the MOV M,A instruction.

The opcode fetch cycle shows the address (2000H) going out over the address bus, and the opcode for the MOV M,A (77H) returning over the data bus. During the write cycle, the address that was stored in the H and L registers goes out from the microprocessor during T_1, and the data to be written from the accumulator goes out during T_2–T_3.

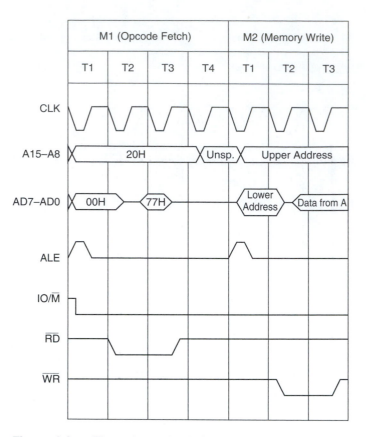

Figure 4-6 ■ The write cycle timing diagram

By the way, this is a case where if the address and data bus were not multiplexed, the microprocessor could overlap the sending out of the data with the sending out of the address, and the instruction could be done faster. But, due to the multiplexed bus, the address must be sent out first, and then the data is sent out after the address has time to be received.

I/O Read

The **I/O read** cycle occurs when an IN instruction is executed by the microprocessor, and during the I/O read cycle, data is read in from an I/O device. In the case of the instruction IN PORT, there are three machine cycles: the opcode fetch cycle to retrieve the opcode; a memory read cycle to get the 8-bit port address into the microprocessor from memory; and, finally, an I/O read cycle where data is read in from the device which has the port address that is sent out during the I/O read cycle. These three cycles take ten T-states.

Figure 4-7 shows the timing of an IN 80H where 80H is the port address of the device being read.

The opcode fetch cycle shows the address of the instruction (2000H) going out over the address bus, and the opcode (DBH) for the IN instruction returning on the data bus. The memory read cycle (M2) displays the address of the second byte of the instruction (20001) going out over the address bus, and the port address (80H) returning on the data bus.

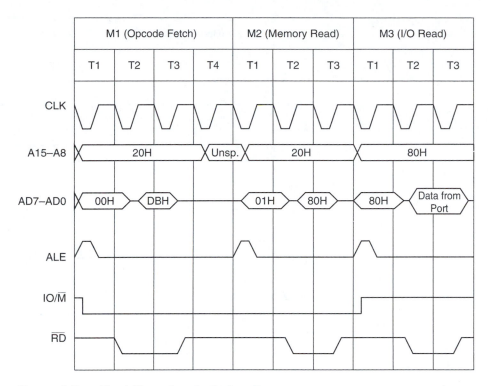

Figure 4-7 ■ The I/O read cycle timing diagram

During the I/O read cycle, the port address of the device being read is sent out over the address bus during T_1—both the upper 8 bits of the address bus and the lower 8 bits carry the identical 8-bit port address, and the data from the input device is returned on the data bus during T_2–T_3.

This is a feature fairly unique to the 8085 microprocessor in that the 8-bit port address of an I/O device is sent out on both the upper and lower portions of the address bus. Many microprocessors don't do this. I suppose this feature is meant to make the interfacing of I/O devices simpler since when designing an interface you can take the I/O address off of the upper 8 bits of the address bus or the lower 8 bits. I'm not sure if that's really a benefit, but it sounds good.

It should also be mentioned at this point that when using peripheral-mapped I/O, you're restricted to only 8 bits for the port address of an I/O device.

I/O Write

The **I/O write** cycle writes data out from the accumulator in the micro-processor to the I/O device specified by the port address.

The OUT Port command has three machine cycles: first, an opcode fetch cycle, then a memory read to get the port address, and then an I/O write cycle. The OUT command writes the data stored in the accumulator over the data bus to the device whose port number was sent out over the address bus. The three machine cycles that make up the OUT Port command are composed of ten T-states.

Figure 4-8 shows the timing of an OUT 84H command. 84H is the port address of the output device.

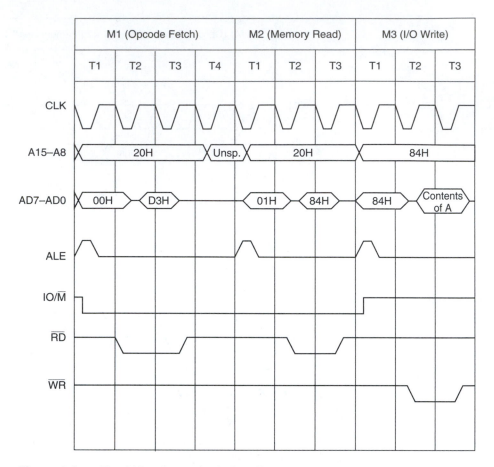

Figure 4-8 ■ The I/O write cycle timing diagram

The opcode fetch cycle sends the address of the instruction (2000H) over the address bus, while the opcode (D3H) for the OUT instruction is returned on the data bus. The second machine cycle shows the address of the second byte of the instruction (2001H) going out over the address bus, with the port address (84H) returning over the data bus.

It can be seen in the I/O write machine cycle that the port address (84H) is sent out over both the upper and lower parts of the address bus, just like the I/O read cycle. Anytime the microprocessor addresses an I/O device, the port number of the device being read or written is sent out over both parts of the address bus.

In Figure 4-8, it can be seen that the data in the accumulator is being written to the output device and goes out over the data bus during T_2–T_3, after the port number has been sent over the two parts of the address bus during T_1 of the I/O write cycle.

Interrupt Acknowledge

The **interrupt acknowledge** machine cycle is a special machine cycle that is used in place of the opcode fetch cycle in the RST (restart) instruction. It is the same as an opcode fetch except that it sends out an $\overline{INTA}$ signal instead of an $\overline{RD}$ signal, and the status lines $IO/\overline{M}$, S_0, and S_1 are 111 instead of 011. In addition, it is six T-states long, not four T-states like the opcode fetch cycle.

4-2 Memory Interfacing

Microprocessor systems must communicate or interface with memory. ROM is often used to store the program that is running the microprocessor, while RAM is used for temporary storage for calculations or other operations done in the microprocessor. Sometimes the program actually resides in RAM, and in those cases, no ROM is necessary. (This is not advised, however, due to the volatility of storing the program in RAM—it's too easy to lose it.) Or, as in some embedded microprocessor systems, only ROM is necessary—no RAM is needed. In other words, in these embedded systems, no operations are being performed that require RAM.

Interfacing to ROM and RAM

In order to communicate with RAM and ROM, an interface is necessary between the RAM and ROM memories, the address bus, the data bus, and the control bus from the microprocessor. The address bus interface is necessary in order to send the address of the memory location that is to be read or written to the memory.

The data bus interface is necessary in order to send the data back (or out in the case of a write) from the memory to the microprocessor.

The control bus interface, on the other hand, is necessary to enable the memory chip(s), so that when an address comes down the address bus that's intended for the memory chip, the memory chip knows whether it is a memory read or a memory write.

Figure 4-9 shows a typical, functional 8085 microprocessor system with both ROM and RAM memory.

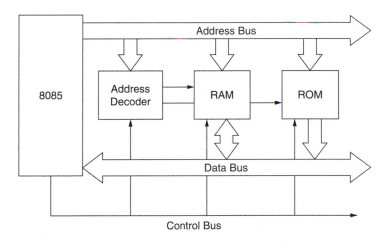

Figure 4-9 ■ The 8085 microprocessor system with RAM and ROM

As shown in Figure 4-9, both RAM and ROM have an interface with the address bus, the data bus, and the control bus. Also shown is the address decoder. This decoder determines which memory chip to enable based upon the memory address that comes over the address bus. In most microprocessor systems, the ROM addresses are the lower ones in the address spectrum, and the RAM addresses are the higher ones. Which range of memory addresses is

ROM versus which range of memory addresses is RAM is particular to each microprocessor system and depends upon how you wire the system. This is called the *memory map* or *memory spectrum*.

The Memory Map

A **memory map** shows all of the possible addresses in a microprocessor system and what they are assigned to. Figure 4-10 presents a simplified, typical memory map for a microprocessor system.

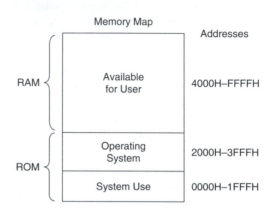

Figure 4-10 ■ A typical memory map

Figure 4-10 shows the lowest range of the address spectrum is ROM and is reserved for system use: 0000H–1FFFH. This includes things like interrupt vector locations, boot-up programs, and other system information. The next ROM area is 2000H–3FFFH and holds the operating system for this microprocessor system. The RAM area—from 4000H–FFFFH—is for user use, or for application programs.

This is a typical type of memory map. Obviously, the sizes and uses of the different areas depend upon the design of the microprocessor system.

The memory map is as large as the address spectrum of the microprocessor. The 8085, with its 16 address bits, can address 2^{16} memory locations. That amounts to 65,536 memory locations, 8 bits wide. In hexadecimal, 65,536 is equal to FFFFH. FFFFH is 16 bits, or in binary is 1111111111111111.

Memory Address Decoding

Address decoding is needed in microprocessor systems in order to select the correct memory chip where the contents of the address being requested resides. In the simplest case, if a memory chip existed that was 64K rows by 8 bits, then the entire 16-bit address bus would go directly to the one and only memory chip. However, this is not the case. Memory chips come in a variety of sizes, but not generally that large. Also, the ROM and RAM chips are separate.

Take the case where the memory chips that need to be addressed by an 8085 microprocessor are each 16K × 8. (This is 16K rows by 8 bits.) To address all 16K rows on each of the chips requires 14 of the address lines. Therefore, the lower 14 address lines would go to each of the four chips required to make up a 64K memory system.

Figure 4-11 shows this memory system composed of four 16K × 8 memory chips, for an 8085 microprocessor system.

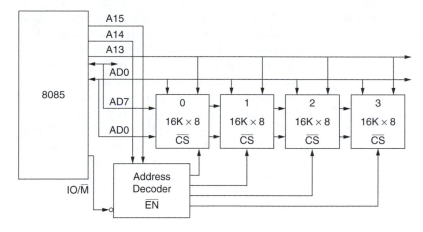

Figure 4-11 ■ An 8085 memory system

The remaining two address lines, the highest order ones, are used to decode which of the four memory chips is to be enabled. In this fashion, all four chips see the lower 14 bits of the address, but only one of the chips is actually enabled, which is determined by the entire 16-bit address.

Address lines A14 and A15 go into the address decoder, and for each address sent down the address bus, only one of the four outputs goes active, enabling one of the memory chips via the chip select (CS). AD0–AD7, shown in Figure 4-11, is the data bus that all of the chips are connected to in order to return data, or get data to be written.

The address decoder circuit in Figure 4-11 needs to activate one out of four outputs depending upon the two input bits. Ways to implement an address decoder will be discussed in the next section. The decoder circuit also has an enable ($\overline{EN}$) input that is tied to the IO/$\overline{M}$ line from the 8085. So, when the 8085 is performing a memory operation, the IO/$\overline{M}$ line goes active low, which enables the address decoder circuit. So, the decoder circuit only sends out the chip select signals to the memory chips when the 8085 is performing a memory operation.

Figure 4-12 shows the layout of the full 16 address bits for this particular system.

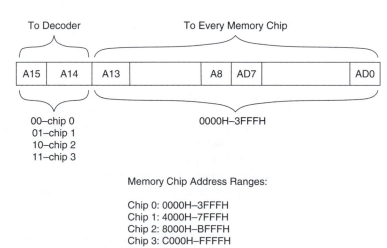

Figure 4-12 ■ The address layout for the memory system

The top two bits pick the chip, and the remaining 14 bits pick the row in that chip. Bits A15 and A14 are sent to the decoder and determine which chip to enable based on the codes shown in Figure 4-12. A13–AD0 are sent to every memory chip in parallel, and only the selected one will put its data onto the data bus. The address ranges for each chip are shown in Figure 4-12.

Figure 4-12 shows four RAM chips, but this addressing scheme can work for any mix of ROM and/or RAM chips, as long as each chip is 16K × 8.

This is the basic method used in addressing memory chips in a microprocessor system. Depending upon the size of the memory chips used, that number of address lines is routed to all of the chips and the remaining address lines are used to decode which chip to enable. As long as the total memory system is always 64K, there will be enough address lines for both the decoder circuit and the memory chips.

Decoder Circuits

Decoder circuits can be made up of individual gates (combinational logic) or they can be decoder or demultiplexer chips. Figure 4-13 shows a decoder for the system from Figure 4-11 made up of individual logic gates.

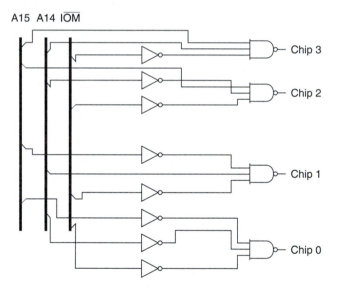

Figure 4-13 ■ A combinational logic decoder circuit

Figure 4-13 shows a decoder circuit composed of four three-input NAND gates and eight inverters. For each value of A15–A14, one and only one of the NAND gates will go active and enable that particular chip, as long as the $IO/\overline{M}$ line is active low. The $IO/\overline{M}$ line going low indicates that the 8085 is performing a memory operation.

Note that on each memory chip, the chip select $(\overline{CS})$ is an active low input. Therefore, when the appropriate NAND gate goes active based on the inputs, this is an active low which is required by the chip select inputs.

Another way to create the decoder circuit for this memory system (Figure 4-11) would be to use a decoder or demultiplexer chip. In this case, a 2–4 decoder chip would be needed.

Figure 4-14 shows the memory decoding circuit for the system in Figure 4-11 using a 2–4 decoder chip.

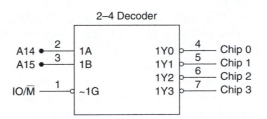

Figure 4-14 ■ The demultiplexor decoder circuit

A14 and A15 are inputted to the 2 to 4 decoder, and based upon the value of these two bits, one and only one of the outputs goes active (active low) when the enable input ($\overline{G}$) goes low from the $IO/\overline{M}$ line. If A15 and A14 are 00, then the output going to chip 0 goes active low. If A15 and A14 are 01, then the output going to chip 1 goes active low. Chip 2 is enabled with a 10 and chip 3 with an 11.

Using decoder chips is usually simpler than using combinational logic for the address decoder in microprocessor systems. However, there may be times when the use of a combinational logic circuit is preferable, but this is usually in smaller or simpler systems.

Example 4.2

▼

Problem: Design a memory system for an 8085 microprocessor system using 8K × 8-bit memory chips.

Solution: Since an 8085 microprocessor can address 64K memory locations, using chips that are 8K each will require eight chips in total.

Also, since each chip has 8K memory words (8 bits wide), 13 of the address lines will need to go to each chip in order to select one of the 8K words on that chip.

This leaves three bits to be used to select the correct memory chip ($2^3 = 8$).

So, using a decoder for the memory address decoder will require a 3–8 decoder. Each of the eight outputs will go to the CS—chip select—for one of the chips.

Figure 4-15 shows this memory system.

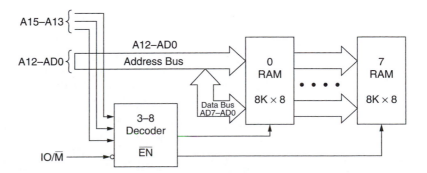

Figure 4-15 ■ Example 4.2—Memory system

In Figure 4-15, the lowest 13 address lines go to each of the eight RAM chips (A12–AD0). Also, the data bus (AD7–AD0) goes to each of the RAM chips in parallel. Note that the data bus shares lines with the low-order 8 bits of the address bus due to being multiplexed.

The 3–8 decoder makes a good address decoder, and using A15–A13 selects one of the eight RAM chips based upon the value of A15–A13. Note that the enable $(\overline{G})$ must be active low from the $IO/\overline{M}$ line for the outputs from the decoder to be active.

4-3 Input/Output Interfacing

Interfacing I/O devices to the 8085 microprocessor is similar to interfacing memory devices, except that I/O devices have two different ways of being interfaced to the system. As discussed in Chapter 2, there is the memory-mapped I/O method of addressing and the peripheral-mapped I/O method.

The Memory-Mapped I/O Method

The **memory-mapped I/O** method of addressing I/O devices means basically including the addresses of the I/O devices within the memory address range and therefore treating them like any other memory location. As such, the programmer can use any memory instructions to read or write the I/O devices. The downside is that part of the address spectrum must be assigned to the I/O devices and is not usable for RAM or ROM.

When using the memory-mapped method, the programmer may not use the input/output instructions IN and OUT of the 8085 since they cause the 8085 control line $IO/\overline{M}$ to go high, indicating an I/O operation and not a memory operation. But, they may use any of the load and store memory instructions, which cause the 8085 control line $IO/\overline{M}$ to go low, signifying a memory operation.

So, in order to interface to I/O devices using the memory-mapped method, the devices must have an address within the memory address range of the 8085 system, and the hardware interfacing circuitry must basically be the same as the memory interfacing circuitry.

Figure 4-16 shows a sample memory map and I/O map of a system using memory-mapped I/O.

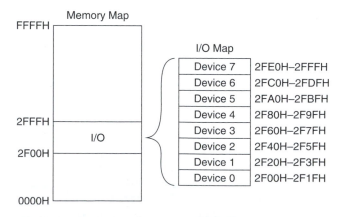

Device 7	2FE0H–2FFFH
Device 6	2FC0H–2FDFH
Device 5	2FA0H–2FBFH
Device 4	2F80H–2F9FH
Device 3	2F60H–2F7FH
Device 2	2F40H–2F5FH
Device 1	2F20H–2F3FH
Device 0	2F00H–2F1FH

Figure 4-16 ■ A sample memory map using memory-mapped I/O

In the system in Figure 4-16, the I/O address range uses the memory addresses of 2F00H to 2FFFH. So, the memory system may not use these addresses in their normal memory address range. Instead, the I/O decoding logic will determine whenever an address is used that is within this range (2F00H–2FFFFH), and the appropriate I/O device will be selected.

In Figure 4-16, each I/O device has a block of addresses, not one specific address. For example, any address from 2F00H–2F1FH will select I/O device 0. The reason for using a block of addresses for each device instead of one specific address is to reduce the number of address lines needed when selecting the device.

In Figure 4-16, to select an I/O device, address lines A15–A8 are needed to select the I/O range, and then only lines A7–A5 are needed to select one of the eight I/O devices. If one specific address was used for each I/O device instead of a block of addresses, you would need to use A7–A0 to select one of the I/O devices. Therefore, using a block of addresses for each device saves on the number of address lines that need to be decoded.

When the address lines A15–A8 have 2FH on them, the instruction is for an I/O device.

Figure 4-17 shows the address lines when addressing an I/O device for this example.

```
A₁₅A₁₄A₁₃A₁₂A₁₁A₁₀ A₉  A₈  A₇  A₆  A₅  A₄  A₃  A₂  A₁  A₀

 0   0   1   0   1   1   1   1   X   X   X  DC DC DC DC DC
```

Figure 4-17 ■ Address lines for memory mapped I/O

In Figure 4-17, bits A15–A8 have 2FH on them to select the I/O address range. Then, bits A7–A5 (shown as X's) are used to select one of the eight I/O devices. Bits A4–A0 are shown as don't cares (DCs) since they are not needed when addressing any of the I/O devices. This is where the savings come in with regards to the number of address lines that need to be decoded. The decoder circuit for this memory-mapped I/O system only needs to use A7–A5 to select one of the eight I/O devices. Address lines A4–A0 are not needed or used in this I/O decoding system.

Figure 4-18 shows the decoding circuit for this memory-mapped I/O system presented in Figure 4-16.

In Figure 4-18, address lines A15–A8 are inputted to an eight-input NAND gate (7430N) and that NAND gate will only have an active output (low) when the address lines are 2FH. This active low signal goes into one of the enable low inputs for the 3–8 decoder (74LS138). The other active low enable for the decoder is connected to the IO/$\overline{\text{M}}$ signal from the 8085, so the decoder chip will be enabled only when IO/$\overline{\text{M}}$ goes low, indicating a memory operation, and when A15–A8 are 2FH, indicating the I/O address block for this example. The other enable input is tied high.

A7–A5 address lines are inputted to the 3–8 decoder, and depending upon the value of these three lines, one of the eight I/O devices will be enabled with an active low signal.

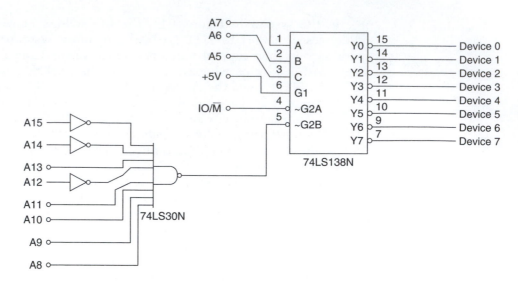

Figure 4-18 ■ A memory-mapped decoding circuit

The Peripheral-Mapped I/O Method

In the **peripheral-mapped I/O** method, the address range for the I/O devices is outside of the address range for the memory of the system—or in other words, it has its own map that is not part of the memory map of the system. Figure 4-19 shows the memory map and the I/O map for a system that uses peripheral-mapped I/O.

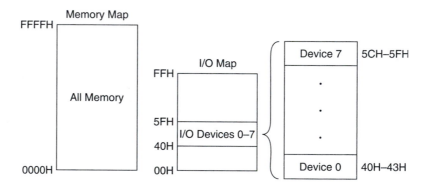

Figure 4-19 ■ The memory map and I/O map for peripheral-mapped I/O

The advantage of this method is that it does not use any of the memory address range. The disadvantage is that only I/O instructions can be used when addressing I/O devices.

When using this method, the I/O address is limited to 8 bits in the 8085, and when an I/O instruction is executed, the I/O 8-bit address is sent out over both the low-order 8 bits of the address bus (AD7–AD0) as well as the high-order 8 bits of the address bus (A15–A8).

Figure 4-19 shows the memory map and I/O map for a system using the peripheral-mapped I/O method.

In the system in Figure 4-19, the memory address spectrum is 0000H–FFFFH and is all usable for memory. The I/O address spectrum is from 00H–FFH, and shown in Figure 4-19 are eight I/O devices that use the I/O address range of 40H–5FH. The rest of the I/O address range is available for other I/O devices. Figure 4-19 shows how the I/O address range 40H–5FH is broken out to access eight devices. Each device uses a block of I/O addresses so that only AD4–AD2 (or A12–A10, if using the upper address lines) need to be decoded in order to select one of the eight I/O devices. The advantage of this decoding scheme is that it requires less logic to decode three address lines as opposed to eight. AD7–AD5 (or A15–A13) are needed to select the I/O address range of 40H–5FH.

Figure 4-20 shows the bit layout of the address lines when addressing these eight I/O devices.

AD7	AD6	AD5	AD4	AD3	AD2	AD1	AD0
0	1	0	X	X	X	DC	DC

Figure 4-20 ■ The address lines layout for peripheral-mapped I/O

Address lines AD7–AD5 are 010 which correspond to the I/O address range of 40H–5FH. Address lines AD4–AD2 thus are used to select one of the eight I/O devices, while address lines AD1–AD0 are don't cares since they aren't used in the decoding circuitry.

Figure 4-21 shows a decoding circuit for this peripheral-mapped I/O system.

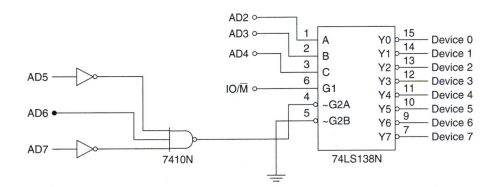

Figure 4-21 ■ A decoding circuit for the peripheral-mapped I/O method

In Figure 4-21, when AD7–AD5 are 010, the active low enable (~G2A) on the 3–8 decoder (74LS138) will be active low, and if it is an I/O operation from the 8085, then IO/$\overline{M}$ will be active high, causing the enable G1 to be active high. Enable ~G2B is tied to ground so it is always active.

Then, AD4–AD2 are used as the three inputs to the 3–8 decoder to select one of the eight I/O devices on the outputs. This is a typical decoding circuit for a peripheral-mapped I/O scheme for the 8085 microprocessor.

Example 4.3

▼

Problem: Design a decoding circuit for a peripheral-mapped I/O system for an 8085 microprocessor. Assume there are 16 I/O devices that need to be addressed, and that the I/O address range for these 16 I/O devices is from 80H–BFH.

Solution: When considering the address range 80H–BFH, note that the first 2 bits remain 10 throughout the entire range. Therefore, AD7 and AD6 can be used as a 10 as part of the decoding circuit.

Then, the next 4 bits are needed to pick one out of 16 I/O devices. So, AD5–AD2 can be used as inputs to a 4–16 decoder whose outputs will select one of the 16 I/O devices. The final 2 bits AD1–AD0 are don't cares and need not be included in the decoding circuit.

A 4–16 decoder would be the easiest way to build this decoding circuit. Figure 4-22 shows a decoding circuit for this example.

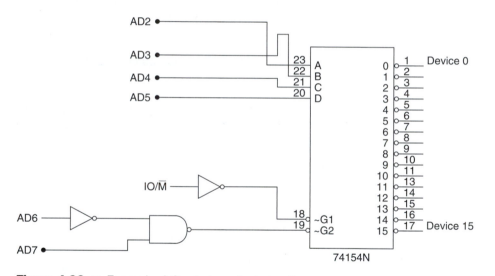

Figure 4-22 ■ Example 4.3—A decoding circuit

In Figure 4-22, AD7 and AD6 are inputted via a NAND gate into one of the enables of the 4–16 decoder (~G2). They are set up so that a 10 will activate the enable (~G2). The other enable (~G1) has the IO/$\overline{M}$ as an input, such that when this control line goes active high—indicating an I/O operation after the inverter—it will put a 0 into the second enable (~G1).

The address lines AD5–AD2 are inputted to the 4–16 decoder chip (74145), and actually select one of the 16 I/O devices. AD1–AD0 are don't cares in this circuit and thus aren't included in the decoding circuit.

▲

Summary

■ The 8085 is an 8-bit processor. The data bus is 8 bits wide, while the address bus is 16 bits wide.

■ The 8085 has a multiplexed bus—between the data bus and the lower 8 bits of the address bus.

- The 8085 chip has 40 pins and can work with a clock up to 3MHz.
- All of the signals of the 8085 can be put into six groups: (1) address bus, (2) data bus, (3) control and status signals, (4) power supply and clock signals, (5) externally initiated signals, and (6) serial I/O ports.
- The 8085's address bus is 16 bits wide and can address 64K bytes of memory. The address bus is split into two groups: AD0–AD7 and A8–A15.
- The group of control and status signals include two control signals ($\overline{RD}$ and $\overline{WR}$), three status signals (IO/$\overline{M}$, S_1, and S_0), and one special symbol—ALE.
- The 8085 has five interrupt signals that can be used to interrupt a program that is executing: INTR, RST7.5, RST6.5, RST5.5, and TRAP. The INTA is the interrupt acknowledgement signal that is an output from the 8085.
- SID (serial input data) and SOD (serial output data) are the I/O ports on the 8085 that are used for inputting and outputting serial data.
- The basic blocks in the 8085's internal architecture are the arithmetic logic unit (ALU), timing and control, the register array, the instruction decoder, interrupt control, and the serial I/O control.
- The arithmetic logic unit (ALU) performs the arithmetic and logical operations.
- The timing and control unit synchronizes all of the microprocessor operations with the clock and generates all of the necessary control signals to the rest of the microprocessor.
- The register array contains all of the programmable registers B, C, D, E, H, and L, as well as two temporary registers, W and X, that are not accessible to general programming use.
- The instruction decoder decodes the instruction and determines the sequence of operations in the microprocessor that must take place.
- The interrupt control unit handles all of the incoming interrupts to the microprocessor, as well as the outgoing interrupt acknowledgement signal.
- The 8085 microprocessor is designed to execute 74 different instructions. Each instruction is divided into two parts: the operation code (opcode) and the operand.
- Each instruction is broken down into machine cycles, and each machine cycle breaks down into clock cycles. These clock cycles are also called T-states.
- The first machine cycle in every instruction is the opcode fetch.
- Machine cycles break down into the following types: opcode fetch, memory read, memory write, I/O read, I/O write, and interrupt acknowledge.
- The ALE (address latch enable) signal goes high during the beginning of every machine cycle.
- Different instructions have different numbers and types of machine cycles in them, and the number of T-states per machine cycle also varies.

- Microprocessor systems often use ROM to store the program that is running the microprocessor, and employ RAM for temporary storage to be used by calculations or other operations done in the microprocessor.

- In order to communicate with RAM and ROM, an interface is necessary between the RAM and ROM memories, and - the address bus, the data bus, and the control bus from the microprocessor.

- A memory map shows all of the possible addresses in a microprocessor system and what they are assigned to.

- Address decoding is needed in microprocessor systems in order to select the correct memory chip within which a requested address resides.

- Decoder circuits can be made up of individual gates, combinational logic, or they can be decoder or demultiplexor chips.

- Interfacing I/O devices to the 8085 microprocessor has two different methods of interfacing: the memory-mapped I/O method, and the peripheral-mapped I/O method.

- The memory-mapped method means basically including the addresses of the I/O devices within the memory address range and therefore treating them like any other memory location.

- In the peripheral-mapped I/O method, the address range for the I/O devices is outside of the address range for the memory of the system.

Questions

1. Why is the 8085 considered an 8-bit microprocessor?
2. What is a multiplexed bus?
3. How much memory can an 8085 microprocessor address and why?
4. Name the six groups of signals in an 8085.
5. Which of the 8085 buses are unilateral?
6. What control line indicates whether it is an I/O or memory operation?
7. What does the ALE signal indicate?
8. Which maskable interrupt is the highest priority?
9. Name the six basic blocks in the 8085 internal architecture.
10. Name the five flags in the ALU and describe what they indicate.
11. Why are the stack pointer and program counter 16 bits?
12. What are the two main parts of any instruction?
13. What are clock cycles also called?
14. What are the six types of machine cycles?
15. What does the opcode fetch machine cycle do?
16. What does the 8085 do when it comes to sending out the 8-bit I/O address?
17. When interfacing to memory, what buses must be interfaced to and why?

18. What is the purpose of a memory map?

19. Describe the general method used in addressing memory chips with regards to the address lines.

20. In what systems would using combinational logic for the address decoder instead of a decoder chip be preferable?

21. How does memory-mapped I/O differ from peripheral-mapped I/O?

22. What is the purpose of having a block of I/O addresses (or memory addresses) apply to each I/O device?

Problems

1. Generate a table showing the status of the control signals $IO/\overline{M}$, $\overline{RD}$, and $\overline{WR}$ for the machine cycles opcode fetch, memory read, memory write, I/O read, and I/O write.

2. What machine cycles and how many T-states total would be in the instruction: MVI B, 55H (move immediate data of 55H into the B register)?

3. What machine cycles and how many T-states total would be in the instruction: JMP 3000H (jump to the address 3000H)?

4. Draw the memory map for a microprocessor system that has ROM from 0000H–2FFFH, where the operating system occupies the upper half of the ROM range, while the lower half is reserved for system use. RAM occupies the rest of the address spectrum and is for applications.

5. Design a memory system for an 8085 that has a full 64K of memory, and uses 16 memory chips of 4K each.

6. Design a combinational logic address decoder circuit for a memory system that uses eight memory chips of 8K each. (You only need to show the address decoder logic.)

7. Show the memory map and I/O map for the memory-mapped I/O system shown in Figure 4-23.

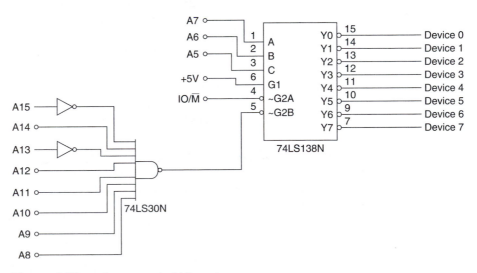

Figure 4-23 ■ An example I/O system

8. Design a peripheral-mapped I/O system using a decoder chip that addresses eight I/O devices that are included in the I/O address range of 80H–9FH.

9. Show the memory map and the I/O map for the I/O system designed in Problem 8.

Laboratory Experiments

Assumptions to Make with Regards to the Following Labs:

- The labs are to be run on an 8085 microprocessor trainer.
- The trainer should have keyboard entry and an LED display.
- The trainer has 2000H as the starting address for user RAM. If your trainer has a different starting address for users, just substitute that base address for 2000H.
- Users should be able to input hexadecimal values to enter a program.
- Hand-assembling is assumed, but if your trainer is connected to a PC and a cross-assembler is available, make the appropriate wording changes in the labs to accommodate this.
- The student can view the contents of memory and registers on the LED display on the trainer. If, however, the student is using a PC interface, make appropriate changes to recognize that the students will be viewing the memory and register contents on the PC.
- The trainer is assumed to have a single-step capability.
- Lab 10 assumes the availability of an 8255 PPI chip and the ability to connect it to the 8085 trainer.
- Lab 11 assumes a basic logic analyzer is available, such as the Tektronix 318/338 Logic Analyzer.
- Lab 12 assumes the availability of a basic PC-based 8085 microprocessor simulator.
- The use of a microprocessor programming sheet and a tracing sheet are assumed. These forms are used when writing a program and when tracing register and memory contents. A sample programming sheet and tracing sheet are attached at the end of this section.

This set of labs is intended to cover a complete 16-week semester where there is one three-hour lab session per week. There are 12 labs, and Labs 9 and 10 are two-week labs, so that covers 14 weeks. Usually the first lab session is not used, and the last lab session is used for make up.

Lab 3: Data Transfer (16-Bit) Instruction Lab

Introduction: This lab will demonstrate the transfer of data to and from 16-bit register pairs.

Objectives: Upon completion of this lab, students will

- Be familiar with the data transfer 16-bit register pair instructions
- Be able to identify the addressing modes: direct addressing, indirect addressing, register addressing, and immediate addressing using 16-bit instructions

- Be able to predict and monitor changes in the contents of registers and memory locations as instructions are executed

Equipment Needed:

- An 8085 microprocessor trainer

Procedure:

1. Hand-assemble the following program using a microprocessor programming sheet. Load it into the trainer starting at location 2000H.

START:	LXI D, 2020H	; load address of 2020H into DE register pair
	MVI A, 55h	; load 55H into A
	STAX D	; store contents of A into location 2020H
	MVI B, 66H	; load 66H into B
	LXI H, 2021H	; load address of 2021 into HL
	MOV M, B	; put contents of B into location 2021H
	XCHG	; exchange HL with DE
	STA 2022H	; store A at location 2022H
	HLT	; halt

2. Single-step the preceding program and record the values of the registers and/or memory that change. Use the tracing sheet to record the values.

3. Hand-assemble the following program using a programming sheet. Load the program into the trainer starting at location 2000H.

START:	LXI D, 2020H	; load address 2020H into DE
	LDAX D	; load A with contents of 2020H
	LHLD 2021H	; load address 2021 into HL
	MOV B, M	; move contents of 2021 into B
	MOV C, A	; move A into C
	MOV A, B	; move B into A
	STAX D	; store A into location 2020H
	MOV M, B	; store B into location 2021H
	HLT	; halt

4. Load location 2020H with DDH and location 2021H with C7H.

5. Single-step the preceding program and record the values of the registers and/or memory that change. Use the tracing sheet to record the values.

Questions:

1. List one example of immediate addressing from this lab using 16-bit instructions.
2. List one example of indirect addressing from this lab using 16-bit instructions.
3. List an example of register addressing from this lab using 16-bit instructions.
4. In the instruction LHLD 2031H what value goes into the H register and what value goes into the L register?
5. Write a brief conclusion for this lab.

Lab 4: Arithmetic Instruction Lab

Introduction: This lab demonstrates use of arithmetic instructions for both 8-bit and 16-bit numbers.

Objectives: Upon completion of this lab, students will

- Be familiar with 8-bit arithmetic instructions
- Understand 16-bit arithmetic instructions
- Be able to predict and monitor changes in the contents of registers and memory locations as instructions are executed

Equipment Needed:

- An 8085 microprocessor trainer

Procedure:

1. Hand-assemble the following program using a programming sheet. Load the program into the trainer starting at address 2000H.

START:	MVI A, 22H	; load 22H into A
	MVI B, 48H	; load 48H into B
	ADD B	; add B to A
	STA 2020H	; store result at location 2020H
	MVI A, 36H	; put 36H into A
	ADI 77H	; add 77H to A
	STA 2021H	; store result at location 2021H
	HLT	; halt

2. Clear locations 2020H and 2021H before starting.

3. Single-step the program and record the changes in registers/memory on the tracing sheet.

4. Hand-assemble the following program using a programming sheet. Load the program into the trainer starting at address 2000H.

START:	MVI A, FDH	; load FDH into A
	MVI B, 32H	; load 32H into B
	SUB B	; subtract B from A
	STA 2020H	; store results at location 2020H
	MVI A, A7H	; load A7H into A
	SUI 42H	; subtract 34H from A
	STA 2021H	; store the results at location 2021H
	HLT	; halt

5. Clear locations 2020H and 2021H before starting.

6. Single-step the program and record the changes in registers/memory on the tracing sheet.

7. Hand-assemble the following program. Record it on a programming sheet.

START:	LHLD 4376H	; load 4376H into A
	LXI B, 1234H	; load 1234H into BC
	DAD B	; adds BC to HL
	SHLD 2020H	; store L at 2020H and H at 2021H
	MVI A, 56H	; load 56H into A
	MVI B, 33H	; put 33H into B
	ADC B	; add B to A with carry
	STA 2022H	; store A at location 2022H
	HLT	; halt

8. Clear locations 2020H, 2021H, and 2022H before starting.
9. Single-step the program and record changes in registers/memory on the tracing sheet.

Questions:

1. What is in the accumulator after the following instructions are executed? What flags are set?

   MVI A, 34H
   MVI B, 76H
   ADD B

2. What is in the accumulator after the following instructions are executed? What flags are set?

   MVI A, 8AH
   MVI B, 1FH
   SUB B

3. What is in the accumulator after the following instructions are executed? What flags are set?

   MVI A, AFH
   ADI 44H

4. Write a brief conclusion for this lab.

5

Programming the 8085—Basic Commands

Objectives:

Upon completion of this chapter, you should:

- Be familiar with the programming model of the 8085
- Be able to program the 8085
- Comprehend data transfer operations
- Understand the addressing modes of the 8085 instruction set
- Be familiar with arithmetic operations
- Understand logical operations
- Comprehend branching operations

Key Terms:

- **Register pairs**—Two registers that are set side by side and used as a single register in order to create a register that is double the normal size

- **Immediate addressing**—An addressing mode in which the instructions contain the data

- **Register addressing**—An addressing mode in which the instructions contain the source and destination as registers

- **Direct addressing**—An addressing mode in which the instructions contain the memory address

■ **Indirect addressing**—An addressing mode in which the instructions store the memory address to be used in the instruction in a register pair

Introduction

Programming the 8085 is very similar to programming any microprocessor. As discussed in Chapter 1, a microprocessor is programmed at the assembly language level. This is a low level, where the programmer knows which registers he is using, going down to the bit level. All assembly language programming is comparable, meaning you only need to know the specific instruction set for that particular microprocessor in order to program it.

A programmer who has learned any one assembly language has learned how to program at this low, bit-twiddling level. From that point on, learning any other microprocessor language is a simple look-up process. As stated earlier, there are only so many ways you can abbreviate JUMP or MOVE or STORE, and the basic classes of operations are the same for all microprocessors. The only thing that differs is the number of instructions in that microprocessor's assembly language set.

The basic classes of the commands for the 8085 are transfer operations, arithmetic operations, logic operations, branch operations, and machine control operations.

The 8085 programming model also needs to be considered since it shows the layout of the registers in the 8085. When programming the 8085, you need to know the names of the usable registers and how many bits wide they are.

Figure 5-1 shows the programming model for the 8085.

Accumulator A (8)	Flag Register (8)
B (8)	C (8)
D (8)	E (8)
H (8)	L (8)
Stack Pointer (SP) (16)	
Program Counter (PC) (16)	

Figure 5-1 ■ An 8085 programming model

Shown in Figure 5-1 are the six general-purpose registers: B, C, D, E, H, and L. The programming model also shows how large they are (8 bits). These registers are typically used in any 8085 instruction when a general-purpose register is called for.

Also shown are the special-purpose registers: the accumulator (A), the flag register, the stack pointer (SP), and the program counter (PC).

The accumulator (A) is used in all arithmetic and logical operations. Basically, any instruction that performs an arithmetic or logical operation does so in the ALU, often using the accumulator as the operand. The result of which usually ends up in the accumulator.

The stack pointer (SP) is a 16-bit register that points to the top of the stack, which is an area of memory that the 8085 uses as a push-down stack. (The stack will be discussed in more detail in the next chapter.)

The program counter (PC) is a 16-bit register that holds the memory address of the next instruction to be fetched and executed, while the flag register contains all of the flags used by the 8085.

Figure 5-2 shows the layout of the flag register for the 8085.

D7	D6	D5	D4	D3	D2	D1	D0
S	Z		AC		P		CY

Figure 5-2 ■ The 8085 flag register layout

The flags for the 8085 are

- **S (sign flag)** After the completion of a logical or arithmetic operation, if the highest order bit D7 in the accumulator is set, the sign bit is set.
- **Z (zero flag)** If the result is zero, the zero flag is set.
- **AC (auxiliary carry)** This flag is set when a carry is generated between D3 and D4 during the operation. This flag is only used internally for BCD conversion and is not available for general use by programs.
- **P (parity flag)** This flag is set when after an operation there are an even number of 1's. If there are an odd number of 1's, the parity flag is reset.
- **CY (carry flag)** If the operation performed results in a carry being generated from D7, then this flag is set.

The S, Z, P, and CY flags can be used by the programmer via conditional jumps, calls, and returns to make decisions following certain arithmetic or logical operations. The AC flag is not directly accessible via program instructions.

Another feature shown in the programming model (albeit very obscurely) is that the general-purpose registers that are shown side-by-side can be used as register pairs. **Register pairs** are a common way in microprocessors to get a 16-bit register when the general-purpose registers are only 8 bits. (Or to get a 32-bit register when the general-purpose registers are only 16 bits.) By using two registers together, they can operate as one 16-bit register. In the 8085, the register pairs are BC, DE, and HL (shown side by side in the register model).

When an instruction requires the storing of a 16-bit address, these register pairs can be used. This will become more apparent later in this chapter when 3-byte instructions are covered. This use of register pairs is common among microprocessors whose data bus is half the size of the address bus. If the data bus and the address bus are the same size, then there would be no need to use register pairs since all registers would be able to hold a full memory address.

5-1 Data Transfer Operations

Data transfer operations are instructions that transfer or move data. Data can be moved into a register, into memory, between registers, between a register and memory, between memory and a register, or between an I/O device and the accumulator.

In the 8085 instruction format, the destination is first and the source second (for example, opcode destination, source).

The 8085 data transfer instructions are

- **MVI R, data (8 bits)** Move the data into the register R.
- **MVI M, data (8 bits)** Move the data into the memory location whose address is located in HL register pair.
- **MOV R1,R2** Move the contents of R2 into R1.
- **MOV R,M** Move the contents of the memory location whose address is in HL to register R.
- **MOV M, R** Move the contents of register R to the memory location whose address is located in HL.
- **LXI Rp, data (16 bits)** Load the 16-bit data into a register pair, where Rp is the first register of the pair (for example, B = BC pair).
- **LDA data (16 bits)** Load the contents of the memory location specified by the 16-bit data into the accumulator (A).
- **LDAX Rp** Load the data located at the address in the register pair, where Rp is the first register in the pair, into the accumulator (A).
- **LHLD data (16 bits)** Load the contents of the memory location specified in the data into register L and then load the contents of the next successive memory location into register H.
- **STA data (16 bits)** Store the contents of the accumulator (A) into the memory location specified by the 16 bits of data.
- **STAX Rp** Store the contents of the accumulator (A) into the memory location specified by the register pair starting with Rp.
- **SHLD data (16 bits)** The contents of register L are stored at the memory location specified by the 16 bits of data, and the contents of register H are stored at the next successive memory location.
- **XCHG** The contents of registers H and L are exchanged with the contents of registers D and E.
- **IN port (8 bits)** Read into the accumulator (A) the data from an input device whose 8-bit port address is specified. (Note: The IN and OUT instructions often are grouped with the Machine Control Operation Instructions into the I/O, Machine Control Operations Group, but a better place for them is in the Data Transfer Group since that's what they actually do.)
- **OUT port (8 bits)** Output the contents of the accumulator (A) to the output device specified by the designated 8-bit port address. (See the preceding note.)

This is a complete list of the data transfer instructions in the 8085 instruction set.

Appendix I contains the complete 8085 instruction set with opcode hex values, the number of bytes, number of machine cycles, number of T-states, flags set, and examples.

Example 5.1

▼

Problem: Write a program that loads the contents of memory location 2050H to the B register, and then loads the contents of memory location 2070H to the D register.

Then, output to port 80H the contents of the B register, and output to port 90H the contents of the D register.

Solution:

LDA 2050H—Load the contents of 2050H into the accumulator.
MOV B,A—Move the contents of A into B.
LDA 2070H—Load the contents of 2070H into the accumulator.
MOV D,A—Move the contents of A into D.
OUT 90H—Output the contents of the accumulator (also D register) to port 90H.
MOV A,B—Move the contents of B into A.
OUT 80H—Output the contents of the accumulator to port 80H.

▲

Addressing Modes

After considering the data transfer instruction set, the different modes of addressing should be discussed. Four modes of addressing are used in most microprocessors.

The following addressing modes are the ones used by the 8085 instruction set:

- *Immediate Addressing* Instructions that have the data in the instructions are called immediate addressing mode. An example is MVI A,55H. The data 55H is moved immediately into the register A.

- *Register Addressing* Instructions that have the source and destination as registers that are specified in the instruction are called register addressing mode. An example is MOV B,A. The contents of register A are moved into register B.

- *Direct Addressing* Instructions that have the memory address as part of the instruction are considered direct addressing mode. An example is STA 2020H. The contents of the accumulator (A) are stored at the memory address 2020H specified in the instruction.

- *Indirect Addressing* Instructions that have the memory address to be used in the instruction stored in a register pair are considered indirect addressing mode. An example is MOV M, A. The contents of the accumulator A are moved to the memory location whose address is stored in the register pair HL.

These are the four addressing modes used by the 8085 instruction set. Understanding the different addressing modes helps in learning the different data transfer instructions.

5-2 Arithmetic Operations

The instructions in this group perform various arithmetic operations such as add, subtract, increment, and decrement. In most cases, the operation is performed in the accumulator and the results stay in the accumulator. Also, all of the flags are set according to the operation results.

The arithmetic operation instructions for the 8085 are

- **ADD R** The contents of the register R are added to the accumulator (A).
- **ADD M** The contents of the memory location whose address is stored in the register pair HL are added to the accumulator.
- **ADI data (8 bit)** Add the data in the instruction to the accumulator.
- **ADC R** The contents of the register R and the carry bit are added to the accumulator.
- **ADC M** The contents of the memory location whose address is stored in the register pair HL are added to the accumulator.
- **ACI data (8 bit)** The data in the instruction and the carry bit are added to the accumulator.
- **DAD Rp** The contents of the register pair starting with Rp are added to the contents of the register pair HL. The results are stored in HL.
- **SUB R** The contents of register R are subtracted from the accumulator.
- **SUB M** The contents of the memory location pointed to by the register pair HL are subtracted from the accumulator.
- **SUI data (8 bit)** The data in the instruction is subtracted from the accumulator.
- **SBB R** The contents of register R and the carry flag are both subtracted from the accumulator.
- **SBB M** The contents of the memory location pointed to by HL and the carry flag are both subtracted from the accumulator.
- **SBI data (8 bit)** The data in the instruction and the carry flag are both subtracted from the accumulator.
- **INR R** The contents of register R are incremented by one.
- **INR M** The contents of the memory location pointed to by HL are incremented by one.
- **INX Rp** The contents of the register pair Rp (where Rp is the first register of the pair) are incremented by one.
- **DCR R** The contents of register R are decremented by one.
- **DCR M** The contents of the memory location pointed to by HL are decremented by one.
- **DCX Rp** The contents of the register pair, starting with Rp, are decremented by one.
- **DAA** The 8-bit number in the accumulator is adjusted to form two 4-bit BCD digits.

Example 5.2

▼

Problem: Write a program that adds the contents of memory locations 2020H and 2021H and stores the sum in memory location 2022H. Also, subtract the contents of memory location 2021H from 2020H and store the difference in memory location 2023H.

Solution:

LDA 2020H—Load the contents of memory location 2020H into the accumulator.
MOV B,A—Move the contents of the accumulator into B.
LDA 2021H—Load the contents of memory location 2021H into the accumulator.
MOV C,A—Move the contents of the accumulator into C. (Note: the source—register A—is unchanged.)
ADD B—Add the contents of B to the accumulator.
STA 2022H—Store the contents of the accumulator in location 2022H.
MOV A,B—Move the contents of B into the accumulator.
SUB C—Subtract the contents of C from the accumulator.
STA 2023H—Store the contents of the accumulator in location 2023H.

Programming Note: Usually there are several ways to write a program, and your choice may differ from someone else's approach. Nevertheless, remember that the most important attribute of any program is its "ease of understanding"— being straightforward—not tricky, cute, or complicated—in order to save a few lines of code. Any program is going to be read and reread many more times than written, so it's most important that it be easy to follow and understand. If it uses a few more instructions than someone else's, that doesn't matter. Since programs will be read often during troubleshooting, and when being modified in the future, having them be easy to read is key.

▲

5-3 Logic Operations

The 8085 logical operation instructions perform AND, OR, XOR, NOT, and rotate operations. As in the arithmetic operations, the accumulator is usually where the operation is performed, and the result is left in the accumulator.

The 8085 logical operation instructions are

- **ANA R** The contents of register R are ANDed with the accumulator (A).
- **ANA M** The contents of the memory location pointed to by HL are ANDed with the accumulator.
- **ANI data (8 bit)** The data is ANDed with the accumulator.
- **ORA R** The contents of register R are OR'd with the accumulator.
- **ORA M** The contents of the memory location pointed to by HL are OR'd with the accumulator.
- **ORI data (8 bit)** The data in the instruction is OR'd with the accumulator.
- **XRA R** The contents of register R are XOR'd with the accumulator.
- **XRA M** The contents of the memory location pointed to by HL are XOR'd with the accumulator.

- **XRI data (8 bit)** The data in the instruction is XOR'd with the accumulator.

- **CMP R** The contents of the register R are compared to the contents of the accumulator, but the contents of the accumulator are not changed. The flags are set based upon the compare: the Z flag is set to 1 if (A)=(R), and the CY flag is set to 1 if (A)<(R). If (A)>(R), both the Z flag and the CY flag are reset. (The parentheses around a register indicate "the contents of.")

- **CMP M** The contents of the memory location pointed to by HL are compared to the contents of the accumulator, but the contents of the accumulator are not changed. The flags are set based upon the compare: the Z flag is set to 1 if (A)=((H)(L)), and the CY flag is set to 1 if (A)<((H)(L)). If (A)>((H)(L)), both the Z flag and the CY flag are reset.

- **CPI data (8 bit)** The data in the instruction is compared to the accumulator and the contents of the accumulator are not changed. The flags set are the Z flag, set to 1 if (A)=data, and the CY flag, set to 1 if (A)<data. If (A)>data, both the Z flag and the CY flag are reset.

- **RLC** The contents of the accumulator are rotated left 1 bit position. The low-order bit and the CY flag are both set to the value shifted out of the high-order bit position. (See Figure 5-3.)

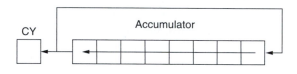

Figure 5-3 ■ An RLC instruction

- **RRC** The contents of the accumulator are rotated right 1 bit position. The high-order bit and the CY flag are both set to the value shifted out of the low-order bit position. (See Figure 5-4.)

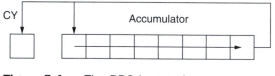

Figure 5-4 ■ The RRC instruction

- **RAL** The contents of the accumulator are rotated left 1 bit position through the CY flag. The low-order bit is set equal to the value of the CY flag, and the CY flag is set equal to the value shifted out of the high-order bit. (See Figure 5-5.)

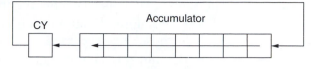

Figure 5-5 ■ The RAL instruction

■ **RAR** The contents of the accumulator are rotated right 1 bit position through the CY flag. The high-order bit is set equal to the value of the CY flag and the CY flag is set equal to the value shifted out of the low-order bit. (See Figure 5-6.)

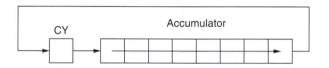

Figure 5-6 ■ The RAR instruction

■ **CMA** The contents of the accumulator are complemented.
■ **CMC** The CY flag is complemented.
■ **STC** The CY flag is set to 1.

Example 5.3

Problem: Write a program that reads in a word from port 85H, saves it at location 201FH, then rotates it left 1 bit position (not through carry flag), ANDs it with 55H, and stores the result in memory location 2020H.

Solution:

IN 85H—Read the data at port 85H into the accumulator.
STA 201FH—Store the contents of the accumulator at location 201FH.
RLC—Rotate the accumulator 1 bit to the left, and the high-order bit goes to both the low-order bit and the CY flag.
ANI 55H—AND the contents of the accumulator with the data 55H.
STA 2020H—Store the contents of the accumulator at location 2020H.

5-4 Branch Operations

The branch operation instructions perform jumps—both unconditionally and conditionally—as well as calls, returns, and restarts.

The conditional jump instructions are based upon the Z flag, the CY flag, the P flag, and the S flag.

The call and return instructions are used in conjunction with subroutines. (Subroutines will be discussed more in the next chapter.)

Restart instructions are used in conjunction with interrupts.

The 8085 branch operation instructions are

- **JMP address (16 bit)** Program control is transferred to the instruction at the address specified.
- **JC address (16 bit)** Program control is transferred to the specified address if the CY flag is set.
- **JNC address (16 bit)** Program control is transferred to the specified address if the CY flag is not set.
- **JZ address (16 bit)** Program control is transferred to the specified address if the Z flag is set.
- **JNZ address (16 bit)** Program control is transferred to the specified address if the Z flag is not set.
- **JP address (16 bit)** Program control is transferred to the specified address if the S flag is not set. (This indicates a plus or positive value.)
- **JM address (16 bit)** Program control is transferred to the specified address if the S flag is set. (This indicates a minus or negative value.)
- **JPE address (16 bit)** Program control is transferred to the specified address if the P flag is set, indicating even parity.
- **JPO address (16 bit)** Program control is transferred to the specified address if the P flag is not set, indicating odd parity.
- **CALL address (16 bit)** Program control is transferred to the specified address, and the address of the instruction following the CALL instruction is stored in the top of the stack. (The stack will be discussed in detail in the next chapter.)
- **CZ address (16 bit)** If the Z flag is set, program control is transferred to the specified address, and the next address after the CZ instruction is stored in the stack.
- **CNZ address (16 bit)** If the Z flag is not set, program control is transferred to the specified address, and the next address after the CNZ instruction is stored in the stack.
- **CC address (16 bit)** If the CY flag is set, program control is transferred to the specified address, and the next address after the CC instruction is stored in the stack.
- **CNC address (16 bit)** If the CY flag is not set, program control is transferred to the specified address, and the next address after the CNC instruction is stored in the stack.
- **CP address (16 bit)** If the S flag is not set, program control is transferred to the specified address, and the next address after the CP instruction is stored in the stack.
- **CM address (16 bit)** If the S flag is set, program control is transferred to the specified address, and the next address after the CM instruction is stored in the stack.
- **CPO address (16 bit)** If the P flag is not set (odd parity), program control is transferred to the specified address, and the next address after the CPO instruction is stored in the stack.
- **CPE address (16 bit)** If the P flag is set (even parity), program control is transferred to the specified address, and the next address after the CPE instruction is stored in the stack.

- **RET** Program control is transferred to the address that is stored on the top of the stack.
- **RZ** Program control is transferred to the address that is stored on the top of the stack, if the Z flag is set.
- **RNZ** Program control is transferred to the address that is stored on the top of the stack, if the Z flag is not set.
- **RC** Program control is transferred to the address that is stored on the top of the stack, if the CY flag is set.
- **RNC** Program control is transferred to the address that is stored on the top of the stack, if the CY flag is not set.
- **RP** Program control is transferred to the address that is stored on the top of the stack, if the S flag is not set.
- **RM** Program control is transferred to the address that is stored on the top of the stack, if the S flag is set.
- **RPO** Program control is transferred to the address that is stored on the top of the stack, if the P flag is not set (odd parity).
- **RPE** Program control is transferred to the address that is stored on the top of the stack, if the P flag is set (even parity).
- **RST n (where (n = 0–7))** Program control is transferred to one of eight fixed memory locations between 0000H and 0038H. These instructions are used in conjunction with interrupts. Also, the address of the next instruction following the RST instruction is stored on the stack.
- **PCHL** The contents of the registers HL are moved into the program counter (PC).

Example 5.4

▼

Problem: Write a program that reads in the word stored at memory location 3000H, and checks to see if the bit 0 is set. If the bit is set, then transfer program control to address 4040H, but if the bit is not set, then increment a counter (use register B), and go back to reading in the memory location 3000H.

Solution:

    MVI B,00H–Zero out the counter register.
    START: LDA 3000H–Read in the word stored at location 3000H. (START is a Label.)
    CMA–Complement the accumulator. (See the following note.)
    ANI 01H–AND the accumulator with 01H.

Note: If bit 0 was originally set, it would have been complemented to 0, and then when ANDed with 01H, the accumulator would be zero if the bit was set. If the bit 0 was not set, it would have been complemented to a 1, and when ANDed with 01H, the results in the accumulator would not be zero. In this way, it's easy to test for a zero condition which denotes that the bit was set. This method of complementing, ANDing, and then testing for zero is a common way to test for a set bit.

    JZ 4040H–Jump to address 4040H if the bit was set.
    INR B–Increment the counter in register B.

JMP START—Jump back to the start of this routine (if hand-assembling this code, you would put the actual address of the LDA 3000H instruction in place of the word START. If using an assembler, labels such as START may be used and the assembler will put in actual instruction addresses.)

Note: Another way to test for bit 0 being set in a word is to read the word into the accumulator, and then use an RAR instruction, followed by a JC 4040H instruction. This moves bit 0 into the CY flag, and if the bit was set, the CY flag will be set and the jump will be taken.

▲

5-5 Machine Control Operations

The machine control operations group includes instructions to control the stack, enable and disable interrupts, perform a "no operation", halt execution, and control serial I/O.

The 8085 machine control instructions are

- **PUSH Rp** The contents of the register pair starting with Rp are pushed onto the top of the stack. (Note: the stack which is covered in Chapter 6 is 8 bits wide.)
- **PUSH PSW** The contents of the accumulator A and then the contents of the flag register are pushed onto the top of the stack.
- **POP Rp** The contents of the top of the stack are put into the first register Rp of the register pair, and the next word in the stack is put into the second register of the register pair.
- **POP PSW** The contents of the top of the stack are put into the accumulator (A), and the next word in the stack is put into the flag register.
- **XTHL** The contents of the top of the stack are exchanged with the L register, and the contents of the next word in the stack are exchanged with the H register.
- **SPHL** The contents of the register pair HL are copied into the stack pointer (SP).
- **EI** The interrupt enable flip-flop is set and all interrupts are enabled.
- **DI** The interrupt enable flip-flop is reset and all the interrupts except for TRAP are disabled.
- **HLT** The processor is stopped.
- **NOP** No operation is performed. (This instruction is often used in timing delay loops.)
- **RIM** This instruction does several things: it reads in the status of the interrupt masks, the status of the interrupts, and the serial input data. (See Figure 5-7.)
- **SIM** This instruction is also a multipurpose instruction. It sets the interrupt masks for RST 7.5, 6.5, and 5.5, resets the RST 7.5 flip-flop, and also sets the serial output data. The accumulator is set up prior to this instruction in order to set the appropriate bits. (See Figure 5-8.)

Accumulator After the RIM Instruction

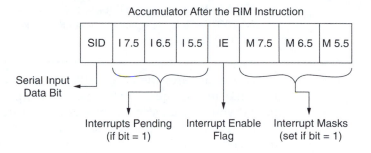

| SID | I 7.5 | I 6.5 | I 5.5 | IE | M 7.5 | M 6.5 | M 5.5 |

Serial Input
Data Bit

Interrupts Pending
(if bit = 1)

Interrupt Enable
Flag

Interrupt Masks
(set if bit = 1)

Figure 5-7 ■ The RIM instruction accumulator

Accumulator Set Up for the SIM Instruction

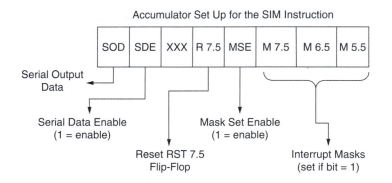

| SOD | SDE | XXX | R 7.5 | MSE | M 7.5 | M 6.5 | M 5.5 |

Serial Output
Data

Serial Data Enable
(1 = enable)

Reset RST 7.5
Flip-Flop

Mask Set Enable
(1 = enable)

Interrupt Masks
(set if bit = 1)

Figure 5-8 ■ The SIM instruction accumulator

Example 5.5

▼

Problem: Write a program that reads in data from input ports 85H and 86H, saves them on the stack, transfers control to a subroutine (not shown in this code), and upon return from the subroutine, pops the data from the stack and stores the data at memory locations 2006H and 2007H.

Solution:

IN 85H–Read in data from input port 85H.
MOV B,A–Move the results into B.
IN 86H–Read in data from input port 86H.
MOV C,A–Move the results into C.
PUSH B–Push first the B register into the stack, and then the C register into the stack.
CALL SUBR1–Transfer control (or call) subroutine SUBR1. (As in the previous example, if you are hand assembling this routine, you would put the actual address for SUBR1 in the instruction. If using an assembler, labels such as SUBR1 can be used.)
POP B–Pop the contents of the stack into the C register and then the B register.
(This instruction is where the RET instruction in the subroutine SUBR1 will return to.)
MOV A,B–Move contents of the B register to the accumulator.
STA 2006H–Store accumulator contents at memory location 2006H.

MOV A,C—Move contents of C register to the accumulator.
STA 2007H—Store the accumulator at memory location 2007H.
HLT—Halt the processor.

▲

This completes the entire 8085 instruction set. It includes many instructions that are used quite often, and others that are far more obscure—something that is typical of most microprocessor instruction sets.

The important thing is to know how assembly language instructions work in general, and be able to look up those instructions that you need. As mentioned earlier, if you learn how to program one microprocessor at the assembly language level, you can program any microprocessor given a list of its instructions, its programming model, and a head start.

Summary

- Microprocessors are programmed at the assembly language level. This is a low level, where the programmer knows which registers he is using, and which goes down to the bit level.

- The basic classes of the basic commands for the 8085 are transfer operations, arithmetic operations, logic operations, branch operations, and machine control operations.

- The six general-purpose registers in the 8085 are B, C, D, E, H, and L.

- The special-purpose registers in the 8085 include the accumulator (A), the flag register, the stack pointer (SP), and the program counter (PC).

- The S, Z, P, and CY flags are accessible via programming, and therefore can be used by the programmer when making decisions following certain arithmetic or logical operations.

- Data transfer operations are instructions that transfer or move data. Data can be moved into a register, into memory, between registers, between a register and memory, between memory and a register, or between an I/O device and the accumulator.

- There are basically four modes of addressing that are used in most microprocessors: immediate addressing, register addressing, direct addressing, and indirect addressing.

- The instructions in the arithmetic operations group perform various operations such as add, subtract, increment, and decrement.

- The logical operation instructions perform AND, OR, XOR, NOT, and rotate operations.

- The branch operation instructions perform jumps, both unconditionally and conditionally, as well as calls, returns, and restarts.

- The machine control operations group includes instructions to control the stack, enable and disable interrupts, perform no operation, halt, and control serial I/O.

- The important thing is to be able to know how assembly language instructions work in general and then be able to look up the instructions you need.

Questions

1. Name the five basic classes of commands used with the 8085.
2. Name the six general-purpose registers and explain why they are considered general-purpose.
3. What is the function of the accumulator?
4. Why is the stack pointer 16 bits wide?
5. What is the function of the program counter?
6. What does it mean when the parity flag is set?
7. What is the purpose of register pairs?
8. What does the instruction MOV A,M do?
9. What does the instruction LXI B, 2020H do?
10. Why is the operand of the IN and OUT instructions only 8 bits?
11. Give an example of direct addressing.
12. Give an example of indirect addressing.
13. Write the instruction that will add the accumulator and the memory location whose address is stored in HL.
14. What does the instruction SUI 44H do?
15. What instruction increments the contents of the register pair DE?
16. What does the instruction ORA B do?
17. Explain the difference between the RLC and the RAL instructions.
18. List the different conditions that the conditional jump commands can be made on.
19. What does the command PUSH B do?
20. What does the command PUSH PSW do?
21. Why must PUSH and POP instructions always be balanced? (In other words, why must they have the same number of pushes as pops?)
22. What are the functions performed by the RIM instruction?

Problems

1. Draw the programming model for a fictitious microprocessor that has a 16-bit data bus and a 32-bit address bus.
2. Write a program that reads in the contents of memory locations 3040H–3043H, and outputs them to ports 80H–83H.
3. What does the following program do?

```
IN 44H
MOV C,A
IN 45H
MOV B,A
CMA
OUT 80H
MOV A,C
CMA
OUT 81H
```

4. Write a program that reads in the inputs from ports 90H–92H and then stores them in consecutive memory locations starting at 2050H.

5. Write a program that adds the two hexadecimal numbers 30H and 40H and stores the sum in memory location 20F0H.

6. Write a program that subtracts two numbers that are to be inputted from ports 55H and 57H, and stores the difference in memory location 40EEH.

7. Write a program that reads in two consecutive words starting at memory location 20FFH, ORs the two words, ANDs the result with 55H, and then stores the result in memory location 3077H.

8. Write a program that reads in 8 bits of data from port 44H, and then counts the number of 1's in the data and stores that result in memory location 2000H.

9. Write a program that examines a word in memory stored at location 2055H, to see if bit 7 (highest order) is set. If set, increment a counter stored in memory location 2077H.

10. What does the following code do?

```
       MVI B,00H
START: LDA 2021H
       CMA
       ANI 04H
       JZ NEXT
       INR B
       JMP START
```

11. Write a program that XORs two words stored at memory locations 1F00H and 1F01H, and if the result is not zero, transfer control to address 4000H. If the result is zero, have the program output a 0FH on port 66H.

12. Write a program that calls a subroutine named DELAY if the first bit (bit 0) in location 3033H is set—and if not set, it should read in the same word again. Upon return of the subroutine DELAY, it should increment a counter stored at location 40000H and return to the beginning of the program.

13. Write a program that reads in the serial input data, and if the bit is a 1, it disables all interrupts (except TRAP). If the bit is a 0, then send out a 0 on the serial output data.

Laboratory Experiments

Assumptions to Make with Regards to the Following Labs:

- The labs are to be run on an 8085 microprocessor trainer.
- The trainer should have keyboard entry and an LED display.
- The trainer has 2000H as the starting address for user RAM. If your trainer has a different starting address for users, just substitute that base address for 2000H.
- Users should be able to input hexadecimal values to enter a program.

- Hand-assembling is assumed, but if your trainer is connected to a PC and a cross-assembler is available, make the appropriate wording changes in the labs to accommodate this.
- The student can view the contents of memory and registers on the LED display on the trainer. If, however, the student is using a PC interface, make appropriate changes to recognize that the students will be viewing the memory and register contents on the PC.
- The trainer is assumed to have a single-step capability.
- Lab 10 assumes the availability of an 8255 PPI chip and the ability to connect it to the 8085 trainer.
- Lab 11 assumes a basic logic analyzer is available such as the Tektronix 318/338 Logic Analyzer.
- Lab 12 assumes the availability of a basic PC-based 8085 microprocessor simulator.
- The use of a microprocessor programming sheet and a tracing sheet are assumed. These forms are used when writing a program and when tracing register and memory contents. A sample programming sheet and tracing sheet are attached at the end of this section.

This set of labs is intended to cover a complete 16-week semester where there is one three-hour lab session per week. There are 12 labs, and Labs 9 and 10 are two-week labs, so that covers 14 weeks. Usually the first lab session is not used, and the last lab session is used for make up.

Lab 5: Logical Instruction Lab

Introduction: This lab will demonstrate how certain logical instructions operate.
Objectives: Upon completion of this lab, the student will:

- Be familiar with certain logical instructions and how they operate
- Understand the term *bit mask* and how it is used in assembly language
- Be able to analyze the use of logical operations and understand how they affect the data in the registers

Equipment Needed:

- An 8085 microprocessor trainer

Procedure:

1. Hand-assemble the following program. Use the microprocessor programming sheets and then load the final program starting at location 2000H.

    This program will take three words in memory: 2020H–2022H and zero out the highest order bit while leaving the other 7 bits unchanged.

```
START:  LXI H, 2020H        ; load memory address
        MVI B, 7FH          ; load bit mask with MSB zero
        MOV A, M            ; load contents of 2020H into A
        ANA B               ; AND A with the bit mask
        MOV M, A            ; load A into location 2020H
```

```
INX H                    ; increment memory address
MOV A, M                 ; load contents of 2021H into A
ANA B                    ; AND A with the bit mask
MOV M, A                 ; load A into location 2021H
INX  H                   ; increment address
MOV A, M                 ; load contents of 2022H into A
ANA  B                   ; AND A with the bit mask
MOV M, A                 ; load A into location 2022H
HLT                      ; halt
```

2. Load locations 2020H to 2022H with FFH.

3. Single-step the preceding program and record the changes in the registers/memory on the tracing sheet.

4. Hand-assemble the following program and record it on the microprocessor programming sheet. Load the program into the trainer, starting at location 2000H.

   This program will take the contents of memory locations 2040H and 2041H, set bits 3 and 4 in each of them, and then return the words back to memory.

```
START: LXI  H, 2040H      ; load memory address
       MVI  B, 14H        ; set bit mask with bits 3 and 4 set
       MOV  A, M          ; load contents of 2040H into A
       ORA  B             ; OR B with A – this sets bits 3 and 4
       MOV  M, A          ; store word at 2040H
       INX  H             ; increment memory address
       MOV  A, M          ; read in contents of 2041H
       ORA  B             ; OR B with A – this sets bits 3 and 4
       MOV M, A           ; store word at 2041H
       HLT                ; halt
```

5. Load locations 2040H and 2041H with 00H.

6. Single-step the program above and record the changes to registers and memory on the tracing sheet.

7. Hand-assemble the following program and record it on the microprocessor programming sheet. Load the program in the trainer, starting at location 2000H.

   The following program reads in a word from memory and checks to see if bit 7 is a one. If it is, it sets it to a zero and halts. If not, the program continues looking.

```
START: LDA  2020H         ; read in contents of 2020H to A
       CMA                ; complement A
       ANI  80H           ; AND A with 80H
       JNZ  START         ; jump if bit 7 was zero
       ANI  7FH           ; set bit 7 to zero and all other bits
                          ; are unchanged
       HLT                ; halt
```

Note: When hand-assembling the preceding program, on the JNZ START instruction, the actual address of the instruction at the START label is used. If the program is loaded at 2000H, then the jump instruction is JNZ 2000H.

8. Set memory location 2020H to 77H initially.

9. Single-step the preceding program and record the changes in registers and memory on the tracing sheet. Verify that the program keeps looping as long as bit 7 is a zero.

10. Set memory location 2020H to 88H.

11. Single-step the preceding program and record the changes on the tracing sheet. Verify that the program goes through once and then halts.

Questions:

1. What is in the accumulator after the following instructions are executed?

   ```
   MVI  A, 55H
   MVI  B, 22H
   ANA  B
   CMA
   ```

2. What is in the accumulator after the following instructions are executed?

   ```
   MVI  A, 17H
   ORI  55H
   RLC
   ```

3. What is in the accumulator after the following instructions are executed?

   ```
   MVI  A, 70H
   MVI  B, 52H
   XRA  B
   RAR
   ```

4. Write the instructions that will take the contents of memory location 2040H, and exclusive OR it with the contents of memory location 2041H.

5. Write a brief conclusion for this lab.

Lab 6: Conditional Jumps Lab

Introduction: This lab demonstrates how conditional jumps work and how they affect program control.

Objectives: Upon completion of this lab, the student will:

- Be familiar with how conditional jump instructions work
- Gain an understanding of how conditional jumps can affect the flow of a program

Equipment Needed:

- A microprocessor trainer

Procedure:

1. Hand-assemble the following program and use the microprocessor programming sheets. Load the program starting at location 2000H.

   Note: When hand-assembling jumps, you must put the label's actual address in the jump instruction (for example, JNZ 200EH).

The following program reads in the contents of memory location 2020H, and if bit 7 is set, it writes a 1 into memory location 2021H, while if bit 7 is not set, it writes a 0 into memory location 2021H.

```
START:  LDA  2020H          ; put contents of 2020H into A
        CMA                 ; complement A
        ANI  80H            ; AND A with 80H
        JNZ  SET1           ; jump if not zero
        MVI  B, 00H         ; load 0 in B
        JMP  STORE1         ; jump to STORE1
SET1:   MVI  B, 01H         ; load 1 in B
STORE1: LXI  H, 2021H       ; load address
        MOV  M, B           ; put B into 2021H
        HLT                 ; halt
```

2. Load 80H into 2020H and single-step the program. Record the changes in registers and memory on a tracing sheet.

3. Load 00H into 2020H and single-step the program. Record the changes in registers and memory on a tracing sheet.

4. Hand-assemble the following program and use microprocessor programming sheets. Load the program starting at location 2000H in the trainer.

   The program that follows reads in memory locations 2033H and 2034H and compares them. If they are equal, an FFH is written into location 2035H. If 2033H is greater than 2034H, an FFH is written into 2036H, and if 2033H is less than 2034H, an FFH is written into location 2037H.

```
START:  LXI H, 2033H        ; load address
        MOV A, M            ; load contents of 2033H into A
        INX H               ; increment address
        MOV B, M            ; load contents of 2034H into B
        CMP B               ; compare B to A
        JC LESS             ; jump if A less than B
        JZ EQUAL            ; jump if A is equal to B
        LXI H, 2036H        ; A is greater than B – set address
        JMP END             ; jump to end
LESS:   LXI H, 2037H        ; A is less than B – set address
        JMP END             ; jump to end
EQUAL:  LXI H, 2035H        ; A=b, set address
END:    MVI A, FFH          ; load FFH
        MOV M, A            ; store FFH at appropriate location
        HLT                 ; halt
```

5. Set memory location 2033H to 25H, and location 2034H to 20H. Clear locations 2035H–2037H.

6. Single-step the preceding program and record the changes in the registers and memory locations on the tracing sheet. Verify that location 2036H gets set to FFH.

7. Set memory locations 2033H to 47H, and location 2034H to 87H. Clear locations 2035H–2037H.

8. Single-step the preceding program and record the changes in the registers and memory locations on the tracing sheet. Verify that location 2037H gets set to FFH.

9. Set memory locations 2033H to 55H and location 2034H to 55H. Clear locations 2035H–2037H.

10. Single-step the preceding program and record the changes in the registers and memory locations on the tracing sheet. Verify that location 2035H gets set to FFH.

Questions:

1. Explain how the CMP instruction sets the CY and Z flags based upon the results of the compare.

2. Will the conditional jump be taken in the following code? Explain your answer.

   ```
   MVI A, 33H
   ADI 52H
   JM LOOP
   ```

3. Will the conditional jump be taken in the following code? Explain your answer.

   ```
   MVI A, 55H
   ORI 02H
   JPO LOOP
   ```

4. Will the conditional jump be taken in the following code? Explain your answer.

   ```
   MVI A, 77H
   XRI 88H
   JZ LOOP
   ```

5. Write a brief conclusion for this lab.

6

Programming the 8085—Advanced Techniques

Objectives:

Upon completion of this chapter, you should:

- Comprehend continuous looping and conditional looping techniques

- Understand counting with single and double registers

- Be able to calculate the length of a delay routine

- Be familiar with the use of register pairs in the technique of indexing

- Know the 16-bit arithmetic operations of the 8085

- Understand how the stack is implemented in the 8085 and know how to use the stack when writing programs

- Be able to write subroutines and use CALL and RETURN instructions

- Understand the use of conditional CALL and conditional RETURN instructions

- Be familiar with how subroutines can be nested

Key Terms:

- **Looping**—A method of setting up a group of instructions to be executed repeatedly in order to do the same tasks over and over

- **Continuous looping**—A method of repeating a set of instructions within a loop over and over indefinitely

- **Conditional looping**—A set of instructions that is repeated when a certain condition is met, and that is not repeated when the condition is not met

- **Counting**—Use of a register to count events, items, and so on

- **Indexing**—Taking a register pair containing a memory address, and then incrementing the register pair by one to access (read or write) consecutive memory locations

- **16-bit arithmetic operations**—Arithmetic operations that deal with the 16-bit register pairs

- **Stack**—A block of memory locations that is set aside and used for a push-down stack

- **Subroutine**—A routine that can be called or transferred to from any place in a program, which after it has been executed, returns to the location that called the program

- **Nesting subroutines**—When a subroutine calls another subroutine, which then calls a subsequent subroutine, and so on

Introduction

Beyond the basic 8085 instructions, a few advanced techniques are worth covering. These include **looping, counting, indexing, 16-bit arithmetic operations, stacks,** and **subroutines**. No new instructions are used in these techniques. They are simply methods that are carried out using the same instructions covered in the previous chapter.

6-1 Looping

Looping is simply setting up a group of instructions to be executed repeatedly in order to do the same tasks over and over. The process can be grouped into two classes: continuous and conditional.

Continuous Looping

Continuous looping repeats the set of instructions within the loop over and over. This, at first, may not seem very useful, but if several applications want to perform the same set of tasks, they can do so, because when the end is reached, the loop goes back to the beginning and starts again indefinitely. This type of system can fit many applications which may be continuously scanning for inputs and performing tasks based on those inputs. An example of this would be a home security system, which constantly scans all the window and door inputs looking for a break-in. Thus, the entire program would loop from the end back to the beginning and repeat its scanning forever.

Figure 6-1 shows the high-level flow chart of a continuous loop in a home security system.

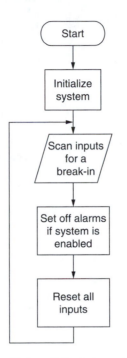

Figure 6-1 ■ A continuous loop example

Continuous loops are implemented with the JMP instruction, which is an unconditional jump to a specific address.

Conditional Looping

An oft-used technique, **conditional looping** is a process in which a set of instructions is repeated if a certain condition is met. If the condition isn't met, however, the instructions are not repeated.

In the 8085, the conditions used for conditional jumps (as listed in the previous chapters) are zero, nonzero, carry, notcarry, even parity, odd parity, plus, and minus. These conditions are stored in the flags that are set or not set based upon the calculations or manipulations that take place in the ALU.

One of the most common uses of conditional looping is to create a routine or program that acts as a delay. A delay routine delays a process for a specified amount of time. Figure 6-2 shows the flow chart for a delay routine that uses the conditional looping technique.

In Figure 6-2, the counter (or register) is initialized to a specific value that will determine the amount of the delay. Then, inside the conditional loop, a certain number of NOPs may or may not be inserted to help take up time in

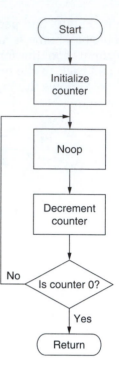

The assembly Language Program for the Routine Above:

```
                MVI B, FFH    ; initialize counter to FF
        START:  NOP           ; No-op
                NOP
                DCR B         ; decrement counter by 1
                JNZ START     ; jump to START if not zero
                RET           ; return
```

Figure 6-2 ■ A conditional looping example

the loop. The counter is decremented, and then a test is made to see if the counter is zero. If not, the loop is repeated. In this example, once the counter reaches zero, the delay routine returns to the calling routine. Figure 6-2 also shows the assembly language program for this routine.

Conditional looping is frequently used in assembly language programming. Whenever a bit is checked to see if it is set, a conditional loop is used. Whenever two values are being compared, a conditional loop is used. The list goes on and on.

6-2 Counting

Counting is simply the act of using a register to count events, items, or whatever the programmer chooses. Usually, a register is initialized to a certain value, and then is counted down or decremented each time the event or condition occurs, using a conditional loop. In Figure 6-2, a delay routine initializes a register, and decrements it by one each time the loop is executed. This is the method used whenever you know how many events are supposed to occur, as in a predetermined delay routine.

It could also be used when you want to do something every tenth time an event occurs, or if you are counting laps and know the number of laps that are to be executed before the race is over.

When you are counting events or items, and you don't know how many there will be, you could use a register to count up from 0. This might be useful for counting how many parts pass by on an assembly line. In this instance, the counter would count upwards.

Another technique might be to use a double register to count down in a delay routine. Often, a single 8-bit register may not give you enough of a delay when they're all 1's, so you can use a double register. Figure 6-3 shows a delay routine that uses a 16-bit register pair to count down.

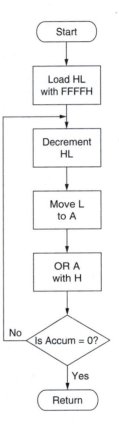

The program for the routine above:

```
        LXI H, FFFFH  ; initialize HL with all 1's
A1:     DCX H         ; decrement HL by 1
        MOV A, L      ; put L into A
        ORA H         ; OR A with H
        JNZ A1        ; jump to A1 if not zero
        RET           : return when zero
```

Figure 6-3 ■ A register pair delay routine

In Figure 6-3, the HL register pair is set to all 1's in order to give a maximum delay. Then, the register pair HL is decremented by 1. In order to check if a 16-bit register pair is all zeros, a technique is used that moves one of the registers of the pair into the accumulator, then OR's the accumulator with the other register of the register pair. This way, if both registers in the pair are zeroes, the result in the accumulator will also be all zeroes. This is a good way of checking to see if a register pair is all zeroes.

It should be mentioned that you can also nest loops in an 8085 assembly language program. Nesting is when you have one loop within another loop. This is both allowed and done often. A simple example occurs, again, in a delay routine. In order to get longer delays, you can put one delay loop within another delay loop. In this way, the entire inner delay loop must be executed each time the outer loop is. This can create large delays when needed. Figure 6-4 shows this type of delay routine.

In Figure 6-4, the C register is used in the inner loop as a counter, while the B register is used as the counter in the outer loop. For each iteration of the

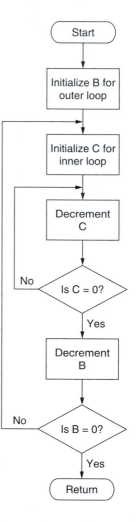

The assembly language program for this routine:

```
            MVI B, FFH    ; initialize B
LOOP1:      MVI C, FFH    ; initialize C
LOOP2:      DCR C         ; decrement C
            JNZ LOOP2     ; xfer if C is not zero
            DCR B         ; decrement B
            JNZ LOOP1     ; xfer if B is not zero
            RET
```

Figure 6-4 ■ A nested loops example

outer (B) loop, the inner loop (C) is executed in total. So, the total delay is the delay time of the inner loop times the number of loops in the outer loop. (This is approximate, since to get an exact delay time you must consider the actual number of cycles in the outer loop plus the total time or number of cycles in the complete inner loop times the number of iterations of the outer loop.) The assembly language program is shown in Figure 6-4. NOPs can also be added into either or both loops if you need to increase the delay of the loops.

This nested loop capability is also used in many other situations and is a common general programming technique.

6-3 Indexing

The technique of **indexing** involves taking a register pair containing a memory address, and then incrementing the register pair by one to access (read or write) consecutive memory locations. In this fashion, you can set up a loop that has an increment register pair in it, and then each time through the loop, the next consecutive memory location will be accessed.

Figure 6-5 shows the flow chart of a program that uses indexing.

The routine in Figure 6-5 moves a block of ten memory words from the starting address to the end address. Ten consecutive words are moved since

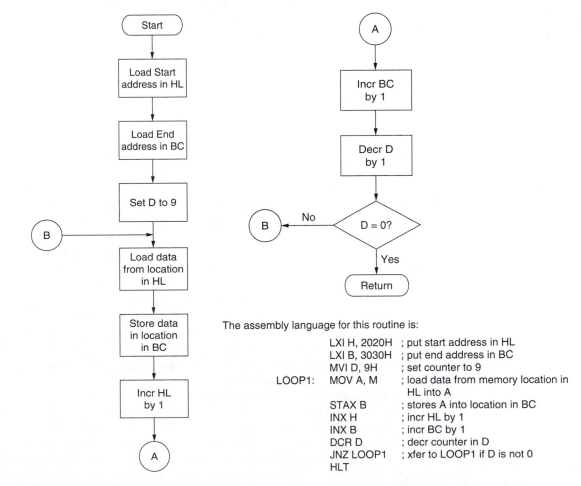

The assembly language for this routine is:

```
        LXI H, 2020H   ; put start address in HL
        LXI B, 3030H   ; put end address in BC
        MVI D, 9H      ; set counter to 9
LOOP1:  MOV A, M       ; load data from memory location in
                         HL into A
        STAX B         ; stores A into location in BC
        INX H          ; incr HL by 1
        INX B          ; incr BC by 1
        DCR D          ; decr counter in D
        JNZ LOOP1      ; xfer to LOOP1 if D is not 0
        HLT
```

Figure 6-5 ■ An indexing example

the counter D is set to 9, which will allow ten iterations through the loop, thus moving ten words: HL+0 to HL+9, to BC+0 to BC+9. D is used as a counter since it is easier to decrement a counter from 9 to 0 and test for 0 than it is to test for HL+9 since you would have to know the address in HL to start. This way, the routine is generic in that it will move ten consecutive words, starting at any location specified in HL, to any location specified in BC.

The code shown in Figure 6-5 implements the flow chart displayed in the same. Note that when reading in the words from memory, a MOV A, M instruction is used, but when storing the data, an STAX B is used. This is because the MOV A, M instruction uses HL as the specified memory address, and the LDAX instruction only allows two register pairs: BC and DE, not HL. So to use HL, the MOV A, M instruction is used, but the STAX B is used when storing the data since the end address is in BC.

6-4 16-Bit Arithmetic Operations

The **arithmetic operations** that use 16 bits in the 8085 are limited. Basically, these are arithmetic operations that deal with the 16-bit register pairs. But in the 8085 instruction set, the only arithmetic operations that deal with register pairs are INX Rp, DCX Rp, and DAD Rp. So, the only arithmetic operations that can be done to the 16-bit register pairs are increment, decrement, and add. This is pretty restrictive. Other microprocessors may have more arithmetic operations you can do with 16-bit words, however. It simply depends upon the microprocessor.

6-5 Stack

The **Stack** in the 8085 is a block of memory locations that is set aside and used for a push-down stack. The stack can be used by programs with the PUSH and POP instructions, and will also be used automatically by the 8085 when it calls a subroutine. When the 8085 calls a subroutine the contents of the program counter are stored in the stack.

The stack functions as a "push-down" stack in that the latest entry is "pushed" onto the top of the stack. As entries are "popped" off the stack, the last entry comes off first.

However, physically, the 8085 implements the stack by initializing the stack pointer (with the LXI SP, it's a 16-bit data instruction), and then when data is put on the stack, it is stored at the memory location one less than the stack pointer, causing the stack pointer to be decremented by 1. In this fashion, the stack won't run out of memory because it physically starts at the highest address (stack pointer) and then works backwards as data is placed on the stack.

The stack, being located in memory, is 8 bits wide, but the PUSH and POP instructions deal only with register pairs, or 16 bits of data. Thus, two 8-bit data words are pushed and popped each time.

Figure 6-6 shows how the stack is implemented in the 8085.

In the first part of Figure 6-6, the stack pointer is initialized to 2300H, and the stack is empty. The B and C registers are loaded with 42D9H. Then, when the PUSH B instruction is executed, the contents of register B are put into the first stack location 22FF, while the contents of register C are placed in the second stack location 22FE. Note that the very first stack location pointed to by the stack pointer when the stack is empty never has data stored in it. Data

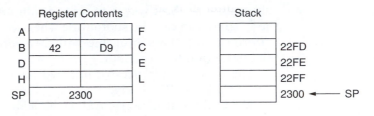

LXI SP, 2300H   ; initialize stack pointer to 2300H
LXI B, 42D9H   ; load 42 in B and D9 in C
PUSH B   ; push first B contents on the stack, then C contents

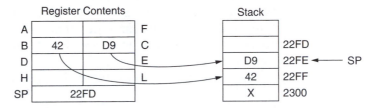

POP B   ; pop contents of stack into BC registers

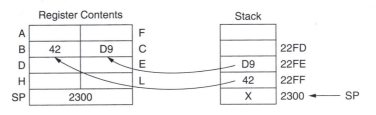

Figure 6-6 ■ The stack implementation

is always stored at the stack pointer minus one. After the PUSH B instruction is executed, the stack pointer (SP) is decremented by two to 22FE.

In the last part of Figure 6-6, when the POP B instruction is executed, the contents of the "top" of the stack are put into register C, while the next word on the stack is placed in register B. After the POP B instruction is executed, the stack pointer is then incremented by two.

The PUSH and POP instructions work on register pairs BC, DE, and HL. But, there is also a PUSH PSW instruction and a POP PSW instruction. These instructions push and pop the accumulator and the flag register. PSW stands for program status word and represents the accumulator with the flag register.

6-6 Subroutines

A **subroutine** is a routine that can be called or transferred to from any place in a program, which after it has been executed, returns to the location that called the program. It is used when a certain task needs to be performed multiple times in a program. By using a subroutine, the code to execute the task needs to be written only once, and then every place in the program that needs that task to be performed can call the subroutine. An example is a time-delay routine. Sometimes programs require a time delay to be performed multiple times; thus, the best way is to write a subroutine for the time delay. This way the program can call the subroutine whenever it needs the time-delay function performed, and the subroutine will then return to the instruction following the instruction that called it.

The 8085 microprocessor has two instructions that are used for subroutines: CALL and RETURN.

The CALL instruction (CALL 16-bit address), when executed, stores the program counter (PC) on the stack, and then transfers control to the address specified in the CALL instruction. Note that the program counter always points to the address of the next instruction to be executed. So, when the CALL instruction is being executed, the PC points to the instruction *following* the CALL instruction.

At the end of the subroutine, the RET instruction "pops" the address off the stack (the address that was "pushed" onto the stack by the CALL instruction), stores it back in the PC, and then transfers control to that address. In this fashion, program control is transferred back to the instruction following the CALL instruction.

Figure 6-7 shows an example of a program that calls a subroutine.

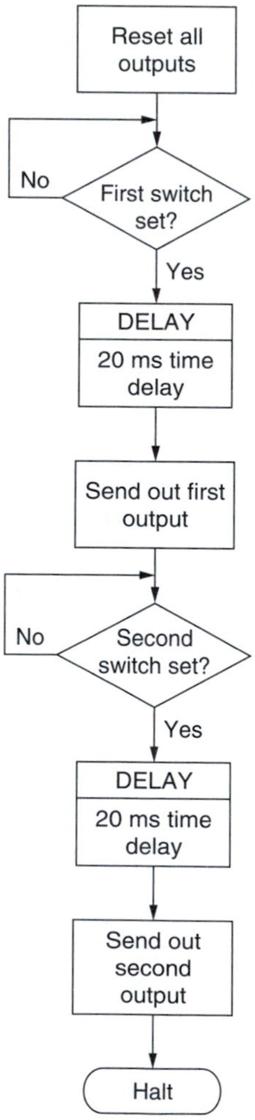

```
              MVI A, 00H
              OUT 08H        ; reset all outputs
LOOP1:  IN 09H               ; read in input switches
              ANI 01H        ; look at first switch
              JZ LOOP1       ; loop back if switch 1 not set
              CALL DELAY     ; go to subroutine to delay 20 ms
              MVI A, 01H     ; set up for first output
              OUT 08H        ; send out first output
LOOP2:  IN 09H               ; read input switches
              ANI 02H        ; look at second switch
              JZ LOOP2       ; loop back if switch 2 not set
              CALL DELAY     ; go to subroutine to delay 20 ms
              MVI A, 02H     ; set up for second output
              OUT 08H        ; send out second output
              HLT

DELAY:  LXI B, 682H          ; load BC register pair with value 682H (1667 decimal)
                                 for a time delay of 20 ms
LOOP:   DCX B                ; decrement BC pair by 1
              MOV A, C       ; move C into A
              ORA B          ; OR B with C
              JNZ LOOP       ; repeat LOOP if BC not zero
              RET            ; return
```

Figure 6-7 ■ A subroutine example

The flow chart in Figure 6-7 shows a program that reads in two switches, one at a time, and then dispatches two outputs when the switches are on. It then calls a time-delay subroutine to provide a 20-ms time delay between seeing that the switch is set and sending out the output for that switch.

Also in Figure 6-7 is the assembly language code for this subroutine example program. It first initializes the outputs by sending out all zeroes on port 08H. It then reads in the input switches on port 09H, and checks the first switch. If the switch was set (1) the code falls through and calls the subroutine DELAY for a 20-ms delay. When the subroutine returns after performing the 20-ms delay, it sends a 1 out to activate the first output on port 08H. This process is repeated for the second switch and the second output.

The DELAY subroutine is a single loop delay routine that uses a register pair instead of a single register in order to get a longer delay. Register pair BC is set to a value of 682H (1667 Decimal) which will make the delay equal to 20 ms assuming the clock speed of the 8085 is 2MHz. The BC pair is decremented, then C is moved to A so that it can be OR'd with B in the next instruction. The results of ORing B and C will only be zero when both registers are zero. If the result is not zero, the loop is repeated. When the register pair reaches zero, the RET instruction is executed, returning program control to the instruction following the CALL instruction.

Time Delay Calculation:

The general equation for calculating the time delay in a loop is

T_D = T-states (in loop) × (Count – 1) × clock period + T-states (in final loop) × clock period

The clock period is $\frac{1}{f}$, where f is the frequency at which the 8085 is being run.

For the delay routine in Figure 6-7 where the clock is 2MHz (which is a clock period of 0.5 us), the delay is

```
DELAY:   LXI B, 682H          ; BC = 682H or 1667D
LOOP:    DCX B                ; 6 T-states
         MOV A, C             ; 4 T-states
         ORA B                ; 4 T-states
         JNZ LOOP             ; 10 T-states if jump taken, 7 T-states if not
```

$T_D = 24 \times 1666 \times 0.5 \times 10^{-6} + 21 \times 0.5 \times 10^{-6}$
$T_D = 20$ ms

Example 6.1
▼

Problem: Design and code a program that operates a drag strip light tree. When the two cars are in place, three yellow lights light up one at a time, with three seconds between them. If either car pulls out before all three yellow lights are lit, a red error light comes on. If both cars stay in place after the three yellow lights come on, a final green "go" light is given.

Solution: The flow chart for the solution is shown in Figure 6-8, along with the assembly language for it in Figure 6-9.

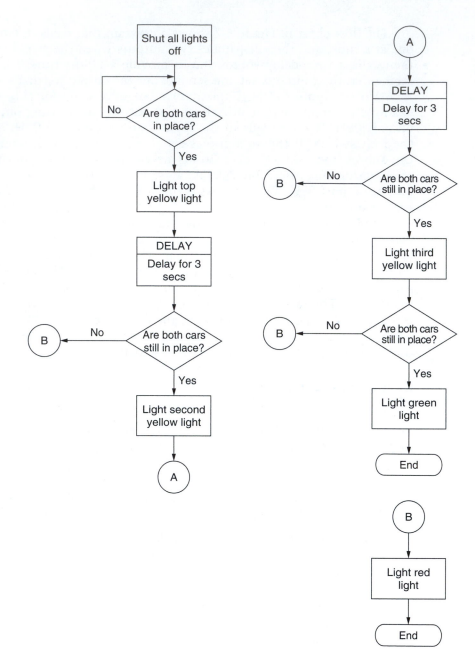

Figure 6-8 ■ Example 6.1—flow chart

Note that a subroutine for the three-second delay is used throughout the program. You could also make the checking of the "cars in place" switches into a subroutine as well. As with all subroutines, the alternative to using a subroutine where the code is written once is to put the code in-line (as in the checking of the switches) every place that the function is needed.

The DELAY subroutine uses a double loop to make the delay longer, while the inner loop is carried out using a register pair. The inner loop is counted down to zero for each value of the outer loop count.

Assembly Language Code for Example 6.1

```
              MVI A, 00H
              OUT 08H          ; turn all lights off
      START:  IN 09H           ; read in "cars in place" switches (1 = in place)
              CMA              ; complement to set 1's to 0's
              ANI 03H          ; look at first two bits (two switches)
              JNZ START        ; loop back if cars not in place
              MVI A, 01H
              OUT 08H          ; turn on first yellow light
              CALL DELAY       ; call delay subroutine
              IN 09H           ; read in switches
              CMA
              ANI 03H          ; look at two switches
              JNZ ERROR        ; go to error if either switch no longer set
              MVI A, 03H
              OUT 08H          ; light second yellow light (keep first one lit)
              CALL DELAY       ; call delay subroutine
              IN 09H           ; read in switches
              CMA
              ANI 03H          ; look at two switches
              JNZ ERROR        ; go to error if either switch no longer set
              MVI A, 07H
              OUT 08H          ; light 3rd yellow light (and 1st and 2nd)
              CALL DELAY       ; call delay subroutine
              IN 09H           ; read in switches
              CMA
              ANI 03H          ; look at two switches
              JNZ ERROR        ; go to error if either switch no longer set
              MVI 0FH
              OUT 08H          ; light green light (keep three yellow lights lit)
              HLT
      ERROR:  MVI A, 10H
              OUT 08H          ; light red light—shut off yellow lights
              HLT
      DELAY:  MVI B, 4H        ; load 4H (4D) into B to give a 3 sec delay
      LOOP2:  LXI D, F424H     ; load F424H (62500D) into DE
      LOOP1:  DCX D            ; decrement DE pair by 1
              MOV A, E         ; put E into A
              ORA D            ; OR D and E
              JNZ LOOP1        ; loop back if DE not zero
              DCR B            ; decrement B by 1
              JNZ LOOP2        ; loop back if not zero
              RET              ; return
```

Figure 6-9 ■ Example 6.1—Assembly language

Time Delay Calculation:

Assume the clock speed is 2MHz. Thus, the clock period is 0.5 us.

```
DELAY:   MVI B, COUNT1      ; load an 8-bit number into B to ultimately give a
                              3-second delay
LOOP2:   LXI D, FFFFH       ; load FFFFH into DE (65535 D) (10 T-states)
LOOP1:   DCX D              ; 6 T-states
         MOV A, E           ; 4 T-states
         ORA  D             ; 4 T-states
         JNZ LOOP1          ; 10/7 T-states
         DCR B              ; 4 T-states
         JNZ LOOP2          ; 10/7 T-states
```

The delay of the inner loop is

$T_D = 24 \times 62499 \times 0.5 \times 10^{-6} + 21 \times 0.5 \times 10^{-6}$
$T_D = 750$ ms
Delay of the two nested loops:
$T_D =$ T-states(outer loop) $\times$ (count $-$ 1)(outer loop) $\times$ clock period $+$ T-states (outer last loop) $\times$ clock period $+$ count (outer loop) $\times$ T_D (inner loop)

So, for Example 6.1:
$T_D = 24 \times 3 \times 0.5 \times 10^{-6} + 21 \times .5 \times 10^{-6} + 4 \times 750 \times 10^{-3}$
$T_D = 3$ seconds

Note that the count of the inner loop was set to 62500D or F424H—not the maximum of FFFFH. This was done in order to make the total delay come out evenly to 3 seconds. If the inner loop was set to FFFFH, then it would require a fraction of loops in the outer loop to get an even 3 secs.(It would require 3.81 loops.) Thus, it was set so that an even number of loops in the outer loop—four—would give an even delay of three seconds.

Also note: in the calculation of the total delay, the T counts of the outer loop are often swamped by the count $\times$ T_D of the inner loop.

So, as an estimate, for a double loop delay routine, one can simply calculate the delay of the inner loop and then multiply that times the count of the outer loop.

▲

6-7 Conditional Call and Return Instructions

To make the calling to (and returning from) subroutines easier, a conditional CALL instruction exists, as well as a conditional RETURN instruction. These conditional instructions are based upon the flags: zero, sign, carry, and parity. When used, the CALL instruction is not executed unless the condition specified is met. The same is true for the RET instruction.

The conditional instructions are

- **CZ address (16-bit)** Program control is transferred to the specified address, while the next address after the CZ instruction is stored in the stack if the Z (zero) flag is set.

- **CNZ address (16-bit)** Program control is transferred to the specified address, and the next address after the CNZ instruction is stored in the stack if the Z (zero) flag is not set.

- **CC address (16-bit)** Program control is transferred to the specified address, while the next address after the CC instruction is stored in the stack if the CY (carry) flag is set.

- **CNC address (16-bit)** Program control is transferred to the specified address and the next address after the CNC instruction is stored in the stack if the CY (carry) flag is not set.

- **CP address (16-bit)** Program control is transferred to the specified address, while the next address after the CP instruction is stored in the stack if the S (sign) flag is not set.

- **CM address (16-bit)** Program control is transferred to the specified address, while the next address after the CM instruction is stored in the stack if the S (sign) flag is set.

- **CPO address (16-bit)** Program control is transferred to the specified address, while the next address after the CPO instruction is stored in the stack if the P (parity) flag is not set. (odd parity)
- **CPE address (16-bit)** Program control is transferred to the specified address, while the next address after the CPE instruction is stored in the stack if the P (parity) flag is set. (even parity)
- **RZ** Program control is transferred to the address that is stored on the top of the stack if the Z (zero) flag is set.
- **RNZ** Program control is transferred to the address that is stored on the top of the stack if the Z (zero) flag is not set.
- **RC** Program control is transferred to the address that is stored on the top of the stack if the CY (carry) flag is set.
- **RNC** Program control is transferred to the address that is stored on the top of the stack if the CY (carry) flag is not set.
- **RP** Program control is transferred to the address that is stored on the top of the stack if the S (sign) flag is not set.
- **RM** Program control is transferred to the address that is stored on the top of the stack if the S (sign) flag is set.
- **RPO** Program control is transferred to the address that is stored on the top of the stack if the P (parity) flag is not set. (odd parity)
- **RPE** Program control is transferred to the address that is stored on the top of the stack if the P (parity) flag is set. (even parity)

Of course, all subroutine calls and returns can be done with the standard CALL and RETURN instructions, but by using the conditional call and return instructions, the programs can be more efficient or use less code.

Figure 6-10 shows an example using conditional call and return instructions. It also shows a program that senses two different temperatures and displays them on the same display when the D/A conversion is complete. In the program, the DISPLAY subroutine is called conditionally if the zero flag is not set (by the CNZ instruction). If the zero flag is set, indicating that the D/A conversion is not complete, the program goes on to sense the next temperature, and does not call the DISPLAY subroutine.

In the DISPLAY subroutine, there is a conditional return—RZ—that returns if the small delay loop is complete. The DISPLAY subroutine has a delay loop in it to allow the LED display time to display that value before getting changed to the next time value.

Note that this program could have just as easily been implemented with regular CALL and RET instructions instead of conditional call and return instructions.

6-8 Nesting Subroutines

A subroutine in 8085 assembly language can call another subroutine, which can call another subroutine, and so on. This is called **nesting subroutines**. The limit on how many subroutines can be nested is determined by how deep the stack is. Every subroutine call stores the return address in the stack. So, as subroutines are nested, the return addresses are pushed down into the stack. Thus, as long as the stack does not overflow, the subroutines can be nested safely.

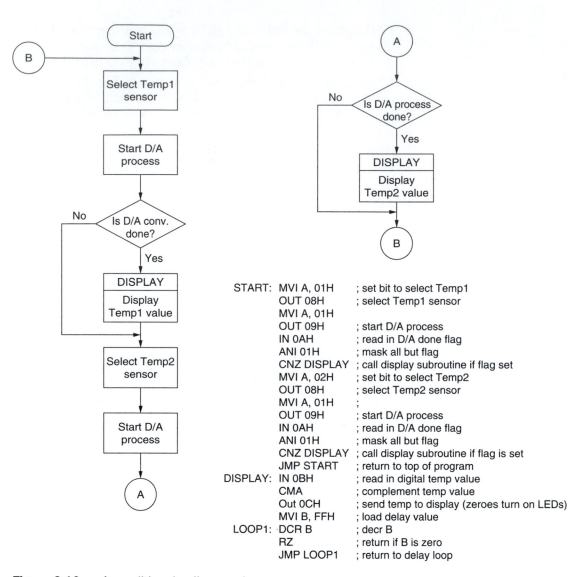

Figure 6-10 ■ A conditional call example

In almost all cases, the number of nested subroutines would never begin to approach in number the available slots in the stack—barring any error condition that fills up the stack with garbage. So, in designing programs, nesting subroutines down several levels would not be a problem.

Figure 6-11 shows the nesting of subroutines graphically. It also shows a main program that calls subroutine 1, which in turn calls subroutine 2. Each time a CALL instruction is executed, the address of the next instruction to be executed when the subroutine returns is stored in the stack. Figure 6-11 shows the contents of the stack when both subroutines have been called but neither return has been executed yet.

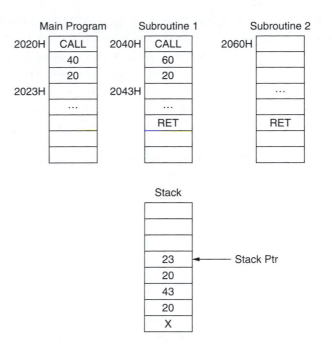

Figure 6-11 ■ A nesting subroutine example

Summary

- Looping is the setting up of a group of instructions to be executed repeatedly in order to do the same tasks over and over.
- Continuous looping repeats the set of instructions within the loop over and over indefinitely.
- Conditional looping is a situation in which a set of instructions is repeated if a certain condition is met, but if that condition is not met, the instructions are not repeated.
- Counting is simply using a register to count events, items, or iterations.
- One counting technique is to use a double register to do a countdown in a delay routine in order to increase the amount of delay.
- In order to get longer delays, you can put one delay loop within another delay loop. In this way, the entire inner delay loop has to be executed each time the outer loop is executed.
- The technique of indexing involves taking a register pair containing a memory address, and then incrementing the register pair by 1 to access (read or write) consecutive memory locations.
- The arithmetic operations that use 16 bits in the 8085 are very limited. Basically, these are arithmetic operations that deal with the register pairs. These instructions are INX Rp , DCX Rp, and DAD Rp.

- The stack in the 8085 is a block of memory locations that is set aside and used for a push-down stack.
- The stack can be used by programs with the PUSH and POP instructions, but is also employed automatically by the 8085 when it calls a subroutine.
- The PUSH and POP instructions work on register pairs BC, DE, and HL. The PUSH PSW stores the accumulator and the flag register on the stack.
- A subroutine is a routine that can be called or transferred to from any place in a program, which after it has been executed, returns to the location that called the program.
- The CALL instruction stores the program counter (PC) on the stack, and then transfers control to the address specified in the CALL instruction.
- The RET instruction "pops" the address off of the stack, stores it back in the PC, and then transfers control to that address.
- Nesting subroutines is when a subroutine can call another subroutine, which can in turn call another subroutine, and so on.
- A conditional CALL instruction exists as well as a conditional RETURN instruction. These conditional instructions are based upon the flags: zero, sign, carry, and parity.
- The limit on how many subroutines can be nested is determined by how deep the stack is. Every subroutine call stores the return address in the stack.

Questions

1. How does continuous looping differ from conditional looping?
2. Give an example of a function other than a delay routine that uses counting with a register.
3. What two methods can you use to make a delay routine last longer?
4. Name two instructions that deal with 16-bit arithmetic operations.
5. How does data get onto the stack?
6. Why is the stack considered a "push-down" stack?
7. Why does the top word in the stack never have data stored in it?
8. What does the instruction PUSH PSW do?
9. What is the advantage of a subroutine?
10. What are the two essential instructions used for subroutines?
11. What happens when an RET instruction is executed?
12. What does the CALL instruction do with the next instruction's address?
13. What are the flags that the CALL and RET instructions can be made conditional on?
14. What is the limiting factor to nesting subroutines?

Problems

1. Write a program that uses a conditional loop to insert a delay of 0.5 ms after reading in an input switch on port 08H, bit 0, and outputting a 0 to light an LED on port 09H, bit 0. (Assume the 8085 cycle time is 2MHz when calculating the delay.)

2. Write a program that increments a counter for every third lap a car goes around a track, and when the counter gets to 10, sound a buzzer (a 1 to bit 0 on port 1AH) and reset the counter. (Assume the lap count comes in on bit 0 on port 1BH.)

3. Write a delay subroutine that uses a register pair to obtain a maximum delay value.

4. Write a delay subroutine that uses two loops of single register delays each to obtain a maximum delay.

5. Write a program that moves a block of 20 words from memory location 2020H to memory location 3020H. (Hint: Use register pairs for the addresses.)

6. Write a program that after initializing the stack pointer, reads in a memory address from locations 2037H and 2038H and then pushes it on the stack.

7. Write a subroutine that when entered, pops off the top address in the stack and then stores it in register pair HL. It then increments the address by two and pushes it back on the stack before returning.

8. Write a program that lights LED1 (port 08H bit 0), calls a delay subroutine that delays for 10 ms, then lights LED2 (port 08H bit 1), calls the same 10-ms delay subroutine, then lights LED3 (port 08H bit 2).

Laboratory Experiments

Assumptions to Make with Regards to the Following Labs:

- The labs are to be run on an 8085 microprocessor trainer.
- The trainer should have keyboard entry and an LED display.
- The trainer has 2000H as the starting address for user RAM. If your trainer has a different starting address for users, just substitute that base address for 2000H.
- Users should be able to input hexadecimal values to enter a program.
- Hand-assembling is assumed, but if your trainer is connected to a PC and a cross-assembler is available, make the appropriate wording changes in the labs to accommodate this.
- The student can view the contents of memory and registers on the LED display on the trainer. If, however, the student is using a PC interface, make appropriate changes to recognize that the students will be viewing the memory and register contents on the PC.

- The trainer is assumed to have a single-step capability.
- Lab 10 assumes the availability of an 8255 PPI chip and the ability to connect it to the 8085 trainer.
- Lab 11 assumes a basic logic analyzer is available, such as the Tektronix 318/338 Logic Analyzer.
- Lab 12 assumes the availability of a basic PC-based 8085 microprocessor simulator.
- The use of a microprocessor programming sheet and a tracing sheet are assumed. These forms are used when writing a program and when tracing register and memory contents. A sample programming sheet and tracing sheet are attached at the end of this section.

This set of labs is intended to cover a complete 16-week semester where there is one three-hour lab session per week. There are 12 labs, and Labs 9 and 10 are two-week labs, so that covers 14 weeks. Usually the first lab session is not used, and the last lab session is used for make up.

Lab 7: Subroutine Lab

Introduction: This lab demonstrates how subroutines work in assembly language. Also, it shows how the stack works in conjunction with subroutines.
Objective: Upon completion of this lab, the student will:

- Understand how subroutines work in 8085 assembly language
- Be familiar with how the stack works when using subroutines
- Be familiar with PUSH and POP instructions

Equipment Needed:

- An 8085 microprocessor trainer

Procedure:

1. Hand-assemble the following program using the microprocessor programming sheets. Store the program in the trainer starting at location 2000H, and load the subroutine starting at location 2050H.

Note: When hand-assembling CALL instructions, you must put the subroutine's actual address into the CALL instruction (for instance, CALL 2050H).

This program calls a subroutine to add the two numbers in the A and B registers and then return with the sum in register C. It uses PUSH PSW (program status word) and POP PSW to save and restore the accumulator and the flag register on the stack when entering the subroutine and restoring it from the stack when returning. Note that the PUSH instruction does not affect the contents of A or the contents of the flag register.

```
START:  LXI  SP, 20C0H       ; initialize the stack pointer
        MVI  B, 23H          ; load 23H into B
        MVI  A, 17H          ; load 17H into A
        CALL ADDN            ; call the ADDN subroutine
        LXI  H, 2070H        ; set up memory address
        MOV  M, C            ; put C into location 2070H
        HLT                  ; halt

ADDN:   PUSH PSW             ; save A and flags on the stack
        ADD  B               ; add B to A
```

```
MOV  C, A                  ; put sum in C
POP  PSW                   ; restore A and flags
RET                        ; return from subroutine
```

2. Load memory location 2070H with 00H.

3. Single-step the program previously listed and record the changes in registers and memory on the tracing sheets. Verify that the flow of the program goes to the subroutine and then back to the instruction following the CALL instruction.

4. Hand-assemble the following program and record the program on the microprocessor programming sheets. Load the main program at location 2000H and load the subroutine at location 2050H.

   This program calls a subroutine that takes the contents of A and stores it at the location specified by HL.

```
START:  LXI  SP, 20C0H      ; set up stack pointer
        MVI  A, 37H         ; load 37H into A
        LXI  H, 2020H       ; load address
        CALL STOREA         ; call subroutine
        HLT                 ; halt

STOREA: MOV  M, A           ; put A into location specified in HL
        RET                 ; return to point of the call +1
```

5. Clear memory location 2020H.

6. Single-step the preceding program and record the changes in registers and memory on the tracing sheet. Verify that the subroutine call is executed and the program then returns to the HLT after the CALL instruction.

7. Hand-assemble the following program using the microprocessor programming sheet. Load the program in the trainer starting at location 2000H, and load the subroutine at location 2050H.

   This program decrements a counter and when the counter reaches zero, it goes to a subroutine to increment another counter by 1. Upon return, when the second counter reaches 3, the program halts. Otherwise, it repeats.

```
START:  LXI  SP, 20C0H      ; set up stack pointer
        MVI  C, 00H         ; set counter to 0
        MVI  D, 2H          ; load count in D

START2: DCR  D              ; decrement A
        CZ   CNT1           ; go to CNT1 subroutine if D is 0
        MVI  A, 03H         ; put 03H into A
        XRA  C              ; XOR A with C
        JNZ  START2         ; repeat if C does not equal 3
        HLT                 ; halt

CNT1:   INR  C              ; increment C
        MVI  D, 2H          ; reset count in D
        RET                 ; return
```

8. Single-step the previous program and record the changes in registers and memory on tracing sheets. Verify that the proper flow of the program is followed.

Questions:

1. Show the contents of the stack after the following code is executed.

   ```
   2000:   LXI  SP, 20C0H
           CALL 2222H
   ```

2. Describe how the stack adds and removes entries.

3. What are the conditional RET instructions that exist?

4. Show the stack and the contents of the stack after the following code is executed.

   ```
   LXI  SP, 20C0H
   LXI  D, 243AH
   PUSH D
   ```

5. Write a brief conclusion for this lab.

Lab 8: Timing and Control Signals and Bus Data Lab

Introduction: This lab uses an oscilloscope to view the control signals of the 8085 microprocessor. It also uses an oscilloscope to look at the data bus and address bus during the execution of a small program.

Objectives: Upon completion of this lab, the student will:

- Be familiar with the ALE, $\overline{RD}$, $\overline{WR}$, and CLK signals of the 8085, and how they occur during a machine cycle
- Gain knowledge of how signals appear on the address bus and the data bus
- Know the overall timing considerations of the 8085

Equipment Needed:

- An 8085 microprocessor trainer
- A dual-channel oscilloscope
- A 40-pin IC test clip

Procedure:

1. Before performing this lab, the student should go back to section 4.1—"The 8085 Microprocessor"—in the text and review how many and what machine cycles are in a typical instruction and how many T-states there are per machine cycle. This will prove very helpful since the student will be looking at timing diagrams and identifying T-states and machine cycles.

2. Hand-assemble the small looping program shown next. Load it into the trainer starting at location 2000H.

   ```
   START:   STA  2045H           ; store A at location 2045H
            JMP  START           ; jump to start
   ```

3. Put a 40-pin IC test clip on the 8085 microprocessor. Be sure it is seated properly on the chip and lined up with the pins.

4. Set up the oscilloscope so that both channels are on 5v/div and the sweep time is set to 1 us/div. Also, set the trigger input to CH2.

5. Connect CH2 to the ALE pin and CH1 to the $\overline{WR}$ pin.

6. Load 5EH into the accumulator (A) and then start the programming going by using the GO command (or whatever command your trainer uses to execute a program).

7. Note that you will have this program running the entire time you are looking at the control signals of the 8085. Thus, all of your analysis will be based upon these two instructions: STA 2045H and JMP 2000H.

8. When observing the two traces on the oscilloscope, you may have to adjust the horizontal hold to get a stable trace.

9. On the ALE trace (CH2), you should see the signal go high at the beginning of each machine cycle. On the $\overline{WR}$ trace (CH1), you should notice only one time where the signal goes low—during the write cycle of the STA instruction.

10. Remember, the STA instruction has four machine cycles: fetch, read, read, and write. The JMP instruction has three machine cycles: fetch, read, and read.

11. So, where the $\overline{WR}$ signal goes low, this indicates where the write cycle of the STA instruction is. Sketch these two traces from the oscilloscope, and then note on the sketch where the machine cycles are and what they are.

12. Switch CH1 to the $\overline{RD}$ pin and leave CH2 on the ALE pin. Sketch the two traces you see on the oscilloscope. Indicate where the machine cycles are and what they are.

13. Switch CH1 to the CLK(out) pin. Sketch the two traces you observe on the oscilloscope.

14. Switch CH1 to the AD0 pin. Note that the first part of the AD0 signal immediately after ALE goes high is the data on the address bus, while the rest of the signal on AD0 is the data on the data bus.

15. Sketch these traces on the oscilloscope.

Questions:

1. Determine the number of machine cycles and identify what machine cycles are in the instruction LXI H, 2020H.

2. Specify what data goes on the data bus and in which machine cycle in the instruction, MVI A, 44H.

3. Determine the number of machine cycles and identify what machine cycles are in the instruction LDA 2020H.

4. What data goes on the data bus for the instruction in Question 3, and in which cycle(s)?

5. Write a brief conclusion for this lab.

7

General-Purpose Support Chips

Objectives:

Upon completion of this chapter, you should:

- Be familiar with five common programmable support chips

- Have a basic understanding of the 8255A Programmable Peripheral Interface (PPI)

- Understand how the 8254 Programmable Interval Timer (PIT) works

- Become familiar with the 8259A Programmable Interrupt Controller (PIC)

- Obtain a basic understanding of the 8237 Direct Memory Access Controller (DMAC)

- Be able to use the 8155 Memory/IO/ Timer chip

Key Terms:

- **Programmable Peripheral Interface (PPI)—** A support chip that sets up multiple parallel input/output ports that are programmable

- **Programmable Interval Timer (PIT)—** A support chip that has multiple counters/timers that can be programmed independently

- **Programmable Interrupt Controller (PIC)**—A support chip designed to manage the interrupts for a microprocessor system

- **Direct Memory Access Controller (DMAC)**—A support chip that transfers data at a high rate from a peripheral device to memory, or vice versa

- **Programmable support chips**—Programmable chips that have a control word that controls the mode the chip is in, as well as the configuration of the chip

Introduction

Many general-purpose programmable support chips are used with a microprocessor when designing microprocessor-based systems. They either interface with peripheral devices or provide functions that can be more efficient when provided by a separate chip. The use of support chips like these in microprocessor system design allows the microprocessor to interface with peripheral devices more easily and perform certain functions more efficiently, freeing up the microprocessor to do other functions.

Five Intel programmable support chips are used quite often with the 8085 microprocessor. These are the **8255A – Programmable Peripheral Interface (PPI)**, the **8254—Programmable Interval Timer (PIT)**, the **8259A—Programmable Interrupt Controller (PIC)**, the **8237 – Direct Memory Access Controller (DMAC)**, and the 8155 Memory/IO/Timer chip.

Since these chips were designed by Intel, they obviously work well with the 8085 microprocessor, but they can also be used with other microprocessors. Each performs useful functions for any microprocessor-based system, and all are programmable, meaning they can be used by any 8-bit microprocessor. Though constructed for general-purpose use, these specific support chips are designed to work with an 8-bit data bus.

Programmable support chips are similar in the manner they are used. Each has a control word that determines the mode the chip is in, as well as the configuration of the chip. This control word is sent out by the microprocessor at the beginning of the program to initialize the support chip. Each chip uses a different control word, but all are programmed via a control word of some sort. As the system designer, you must decide ahead of time how you want the support chips configured and then set up the appropriate control word at the beginning of your program.

7-1 PPI—8255A

The 8255A – Programmable Peripheral Interface (PPI) is a support chip that sets up three parallel input/output ports: A, B, and C. It can be programmed to transfer data under different conditions, from simple I/O to interrupt I/O. Flexible and versatile, it basically has three 8-bit I/O ports that can be programmed either as input or output, or in the case of port C, can be split into two 4-bit ports and programmed separately as input or output.

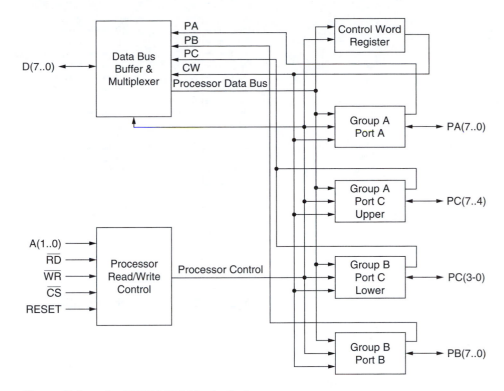

Figure 7-1 ■ An 8255A PPI block diagram

Figure 7-1 shows the block diagram for the 8255A.
In Figure 7-1, the connections are

- **D[7..0]** Data bus
- **A[1..0]** Port address
- **$\overline{RD}$** Read input
- **$\overline{WR}$** Write input
- **$\overline{CS}$** Chip select
- **RESET** Reset input
- **PA[7..0]** Port A—8 bits
- **PC[7..4]** Port C—upper 4 bits
- **PC[3..0]** Port C—lower 4 bits
- **PB[7..0]** Port B—8 bits

These are the connections for the 8255A PPI. The two 8-bit ports—A and B—can be programmed as input or output, while port C can be programmed as two 4-bit ports which can be either input or output.

The data bus from the microprocessor connects into D7 to D0 and is obviously bidirectional.

The $\overline{CS}$ and the A1 and A0 are used in addressing the 8255A. A1 and A0 are used to determine the port address, and the addressing logic employed (based upon the I/O address used for the PPI) is inputted to the $\overline{CS}$.

Figure 7-2 shows the pinout configuration for the 8255A. It also shows the physical pinout connections for the 8255A.

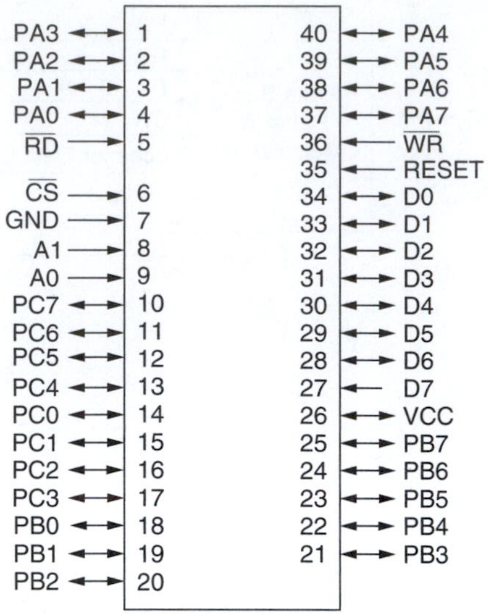

```
PA3 ◄─►  1        40  ◄─► PA4
PA2 ◄─►  2        39  ◄─► PA5
PA1 ◄─►  3        38  ◄─► PA6
PA0 ◄─►  4        37  ◄─► PA7
 RD ──►  5        36  ◄── WR
                  35  ◄── RESET
 CS ──►  6        34  ◄─► D0
GND ──►  7        33  ◄─► D1
 A1 ──►  8        32  ◄─► D2
 A0 ──►  9        31  ◄─► D3
PC7 ◄─► 10        30  ◄─► D4
PC6 ◄─► 11        29  ◄─► D5
PC5 ◄─► 12        28  ◄─► D6
PC4 ◄─► 13        27  ◄── D7
PC0 ◄─► 14        26  ◄─► VCC
PC1 ◄─► 15        25  ◄─► PB7
PC2 ◄─► 16        24  ◄─► PB6
PC3 ◄─► 17        23  ◄─► PB5
PB0 ◄─► 18        22  ◄─► PB4
PB1 ◄─► 19        21  ◄─► PB3
PB2 ◄─► 20
```

Figure 7-2 ■ The pin configuration of the 8255A

In addressing the 8255A, A1 and A0 are used as shown next for the port address:

A1	A0	Hex Address	Port
0	0	0H	A
0	1	1H	B
1	0	2H	C
1	1	3H	Control Word

The $\overline{CS}$ is used for the rest of the I/O address. For example, if the addressing logic shown in Figure 7-3 is used to address an 8255A, then the port addresses would be

- **C0H** Port A
- **C1H** Port B
- **C2H** Port C
- **C3H** Control word

Based upon the addressing logic shown in Figure 7-3, the addressing table becomes

A7	A6	A5	A4	A3	A2	A1	A0	Hex Address	Port
1	1	0	0	0	0	0	0	C0H	A
1	1	0	0	0	0	0	1	C1H	B
1	1	0	0	0	0	1	0	C2H	C
1	1	0	0	0	0	1	1	C3H	Control word

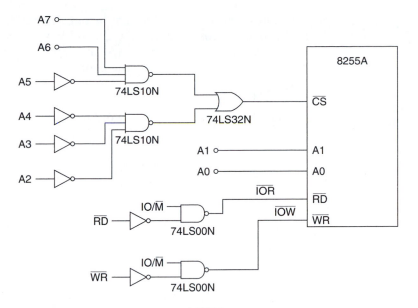

Figure 7-3 ■ Addressing an 8255A

So, when the 8085 addresses the PPI, it uses the address C0H to address port A, C1H to access port B, C2H to access port C, and C3H to access the control word.

The I/O addresses are 8 bits in the 8085. Thus, when an I/O command sends out an address, it sends the 8-bit I/O address out over both the lower 8 bits of the address bus (that is multiplexed with the data bus) and the upper 8 bits of the address bus. So, in Figure 7-3, the A7–A0 can come off of either the lower 8 bits of the address bus, or A15–A8 could also be used since it will have the same I/O address on it.

The $IO/\overline{M}$ signal and the $\overline{RD}$ signal from the 8085 are combined to create an $\overline{IOR}$ signal that is inputted to the 8255A's $\overline{RD}$ connection. This is an I/O read signal. The $IO/\overline{M}$ signal and the $\overline{WR}$ signal from the 8085 are combined to create an $\overline{IOW}$ signal that is connected to the 8255A's $\overline{WR}$ connection. This is an I/O write signal.

When the 8085 executes an OUT command, both the $IO/\overline{M}$ signal and the $\overline{WR}$ signal go active, sending the I/O address out over both A15–A8 and A7–A0.

Similarly, when the 8085 executes an IN command, both the $IO/\overline{M}$ signal and the $\overline{RD}$ signal go active, dispatching the I/O address out over both A15–A8 and A7–A0.

Control Word

When using the 8255A, the designer must first determine the I/O addresses of the ports A, B, C, and the control word. This is determined by the addressing logic connected to the $\overline{CS}$ input and the A1 and A0 inputs (as shown in Figure 7-3).

Next in the program, the control word of the 8255A must be written based upon the configuration you want the 8255A to have, and what mode.

Initially, the 8255A can perform two functions: the I/O function and the bit set/reset (BSR). The I/O function is the one most often used, its function determined by bit D7 in the control word. If D7 = 1, the function is the I/O mode.

The control word has a different layout depending upon bit D7. If D7 = 0, the BSR mode is chosen. When the control word is set to the I/O mode, it configures ports A, B, and C, but when the control word is set to the BSR mode, it only changes port C, leaving the configurations for ports A and B unaffected.

I/O Mode

The layout of the control word for the I/O mode is shown in Figure 7-4.

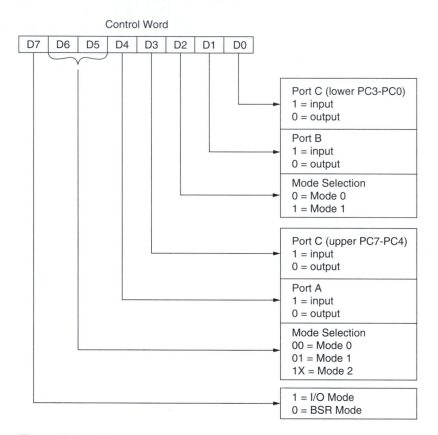

Figure 7-4 ■ The 8255A control word layout—I/O mode

Using the control word, you pick the direction (input or output) for each port: A, B, upper C, and lower C. You also pick the mode (besides the I/O-versus-BSR mode). The I/O mode, for instance, offers three types: mode 0, 1, or 2.

Mode 0 is the simple input/output mode. In mode 0, ports A and B operate as simple input or output 8-bit ports, and C functions as two 4-bit ports—either input or output. The layout of the control word for mode 0 is shown in Figure 7-4.

Mode 1 is the Input or Output Handshaking mode. (Handshaking is the passing back and forth of control signals between the microprocessor and an external device). In this mode, ports A and B function as two 8-bit ports, either input or output. Since ports A and B each use three lines of port C for handshaking signals between the microprocessor and the PPI, the two remaining port C lines can be used for simple I/O functions.

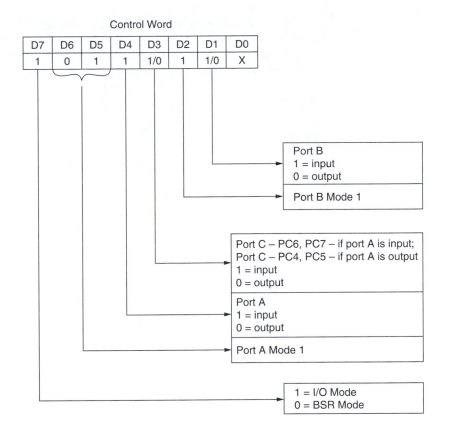

Figure 7-5 ■ Mode 1 control word

The lines used for handshaking in mode 1 vary as to whether the ports are input or output. The control word layout for mode 1 is shown in Figure 7-5.

Port B uses PC0–PC2 for handshaking for both the input and output mode. However, when port A is in the input mode, it uses PC3–PC5 for handshaking, leaving PC6 and PC7 for simple I/O. When port A is in the output mode, it uses PC3, PC6, and PC7 for handshaking, leaving PC4 and PC5 for simple I/O.

The handshaking signals used in mode 1 when either port A or B is in the input mode are

- $\overline{STB}$ **(strobe input)** Indicates from a peripheral device has transmitted a byte of data (port A – PC4, port B – PC2)

- **IBF (input buffer full)** An acknowledgement from the 8255A that it received the byte of data (port A – PC5, port B – PC1)

- **INTR (interrupt request)** An output signal used to interrupt the microprocessor (port A – PC3, port B – PC0)

The handshaking signals used in mode 1 when port A or B is in the output mode are

- $\overline{OBF}$ **(output buffer full)** Output signal from the 8255A to a peripheral indicating data is ready to be read (port A – PC7, port B – PC1)

- $\overline{ACK}$ **(acknowledge)** Input signal from a peripheral indicating it received the data (port A – PC6, port B – PC2)

- **INTR (interrupt request)** Output signal used to interrupt the microprocessor (port A – PC3, port B – PC0)

Mode 2 is the bidirectional data transfer. This mode is used to transmit data between two computers or between a disk controller and a computer. In this mode, port A is bidirectional and uses five lines from port C for handshaking, while port B can be used in either mode 0 or mode 1. The five handshaking signals used by port A when in mode 2 are INTR (PC3), $\overline{OBF}$ (PC7), $\overline{ACK}$ (PC6), $\overline{STB}$ (PC4), and IBF(PC5).

BSR Mode

The BSR mode, which is selected by setting bit D7 to 0 in the control word, is concerned only with port C. This mode allows you to set or reset individual bits in port C. Figure 7-6 shows the layout of the control word for the BSR mode.

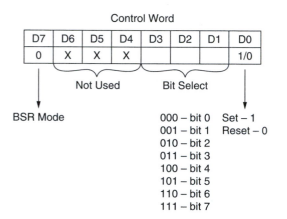

Figure 7-6 ■ The control word for the BSR mode

Remember, when the BSR mode is used, ports A and B remain in the configuration previously set up by the control word in the I/O mode. Thus, they are unaffected by the BSR mode.

In Figure 7-6, having bit D7 set to 0 indicates the BSR mode, bits D6–D4 are unused, and bits D3–D1 are used to select the bit in port C to either set or reset, which is indicated by bit D0. So, with the BSR mode you can set (1) or reset (0) any individual line in port C.

The 8255A PPI is a very common chip that is often used in microprocessor systems. It is very versatile, and provides a good interface for inputting and outputting information to the 8085. It is most often the first support chip selected when designing a system.

7-2 PIT—8254

The 8254 Programmable Interval Timer (PIT) is used for counting and timing. It contains three 16-bit counters that can be programmed independently into one of six different modes. The PIT can be used for time delays, a real-time clock, an event counter, a digital one-shot, a square-wave generator, and a complex waveform generator.

The 8254 PIT is an upgraded version of the 8253 PIT, which is basically the same chip functionally, and is pin-compatible. The 8254 can run at higher clock speeds than the 8253.

As with other programmable support chips, there is a control word that is used to specify the configuration of the three counters.

Figure 7-7 shows a block diagram of the 8254 PIT.

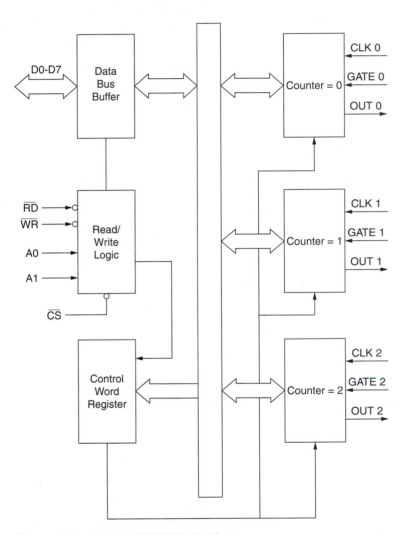

Figure 7-7 ■ An 8254 PIT block diagram

In Figure 7-7, the control word register handles the control word that configures the chip. The read/write logic handles the chip select, the address, and whether it is a read or write. The data bus buffer is the interface to the data bus from the microprocessor. Then there are the three counters 0, 1, and 2. Each counter has two inputs—CLK, GATE, and one output: OUT. How these inputs and outputs are employed depends upon the counter's mode.

Figure 7-8 shows the pinout for the 8254 PIT.

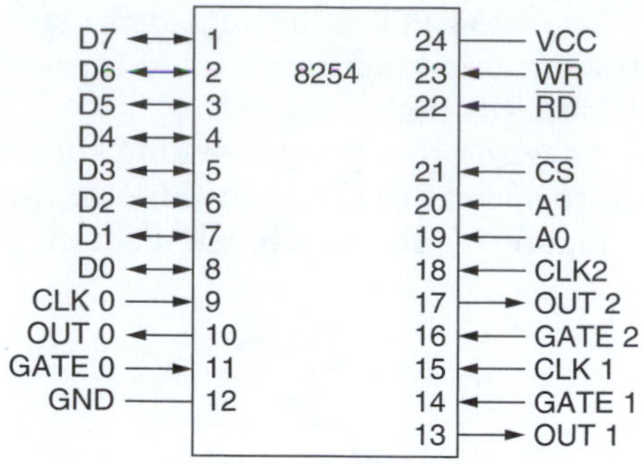

```
        D7 ◄──► 1    ┌──────┐   24 ──── VCC
        D6 ◄──► 2    8254     23 ◄── WR
        D5 ◄──► 3            22 ◄── RD
        D4 ◄──► 4
        D3 ◄──► 5            21 ◄── CS
        D2 ◄──► 6            20 ◄── A1
        D1 ◄──► 7            19 ◄── A0
        D0 ◄──► 8            18 ◄── CLK2
      CLK 0 ──► 9            17 ──► OUT 2
      OUT 0 ◄── 10           16 ◄── GATE 2
     GATE 0 ──► 11           15 ◄── CLK 1
        GND ─── 12           14 ◄── GATE 1
                             13 ──► OUT 1
                    └──────┘
```

Figure 7-8 ■ The 8254 PIT pinout configuration

The counter select bits—A1 and A0—specify which counter and/or control word is being addressed.

A1	A0	Selection
0	0	Counter 0
0	1	Counter 1
1	0	Counter 2
1	1	Control Word

Control Word

The control word for the 8254 is addressed when A1 A0 = 1 1. The layout for the control word is shown in Figure 7-9.

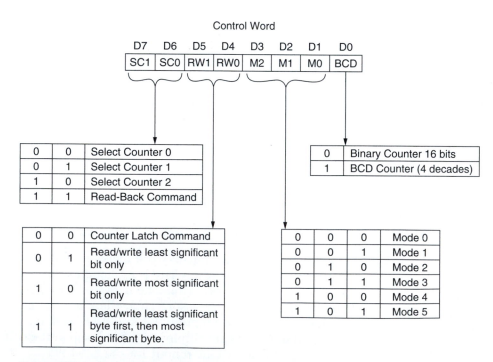

Figure 7-9 ■ The 8254 control word layout

The control word is used to set up each counter. SC1 and SC0 select which counter to be initialized, or choose the read-back command (to be explained later). RW1 and RW0 specify the Counter Latch command, or whether the least significant (or most significant) byte is being read or written. (Each 16-bit counter has 2 bytes in it: the least significant and the most significant.)

M2–M0 specify the mode selected for the counter being addressed. BCD specifies whether the counter is in binary or BCD.

When using the 8254 PIT, you must write the control word to initialize the counter to be used. For every counter employed, you must write the control word to select it and set it up.

Generally, these counters are Initialized to the appropriate mode, the count is loaded (first 1 byte, followed by the other), and then the counter counts down and gives an output when 0 is reached.

Different modes offer variations, but this is the basic way counters are used.

Modes

MODE 0: INTERRUPT ON TERMINAL COUNT

In this mode, the OUT signal is initially low. The count is then loaded into the counter, the counter is decremented with each cycle, and when 0 is reached, the OUT signal goes high. It then stays high until a new count or command word is loaded. Also, the GATE signal may be used to stop the count by setting GATE = 0, and can be continued when the GATE returns to 1.

MODE 1: HARDWARE-RETRIGGERABLE ONE-SHOT

In this mode, the OUT signal is initially high. When the GATE is triggered, the OUT goes low, and when the count reaches 0, the OUT goes high again. Thus, a one-shot signal is generated that is under hardware control (signal on the GATE).

MODE 2: RATE GENERATOR

In this mode, a pulse is generated that is equal to the clock period at a given interval, controlled by the count that is loaded. When the count is loaded, the OUT signal stays high until the count reaches 1, at which point the OUT signal goes low for one clock period. Afterward, the count is reloaded automatically, and the cycle repeats, generating a continuous string of pulses.

MODE 3: SQUARE-WAVE GENERATOR

In this mode, when the count is loaded, the OUT signal is high. The count is then decremented by two with each clock cycle. When the count reaches 0, the OUT signal goes low, and the count is reloaded automatically. This is repeated continuously, generating a square wave on the OUT signal. The count value controls the period of the square wave.

MODE 4: SOFTWARE-TRIGGERED STROBE

In this mode, the OUT signal is initially high, and goes low for one clock period when the count reaches 0. So, one strobe pulse (low) is generated for each count. The count must be reloaded for more strobe signals.

MODE 5: HARDWARE-TRIGGERED STROBE

This mode is similar to mode 4, except the strobe is hardware triggered with a signal on the GATE signal. The OUT signal is initially high. When the GATE signal goes from low to high, the count starts, and when it reaches 0, the OUT signal goes low for one clock period.

READ-BACK COMMAND

The Read-Back command lets you read the count and the status of any of the counters. When the Read-Back command is selected in the control word (SC1 SC0 = 11), each of the counters specified (all three counters can be selected with one control word) is latched, and then the count and/or the status may be read for each counter latched.

The control word layout is different when the Read-Back command is selected. Figure 7-10 shows the layout of the control word in this mode.

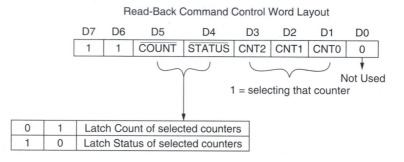

Figure 7-10 ■ The Read-Back command control word layout

Either $\overline{COUNT}$ or $\overline{STATUS}$ must be zero, but not both. You can request either the count or the status to be latched. CNT2–CNT0 specifies the counters to be latched. Once latched, the control word can be used to read each latched counter's count or status. The latched count or status is held until read, or until the counter is reprogrammed. To read the count, you should perform two reads of that counter via the control word. You get the lower byte first, and then a second read is done giving you the higher byte.

When reading the status of a counter, you get back the information in the control word that set the counter up (mode, R/W, BCD/binary) as well as the state of the OUT line.

7-3 PIC—8259A

The Programmable Interrupt Controller – 8259A is designed by Intel to manage the interrupts for the 8085, 8086, and 8088 microprocessors. This support chip works for both 8-bit microprocessors—8085 and 8088 (8-bit data bus)—as well as a 16-bit processor: the 8086. (Note that the 8088 is compatible with the 8086, except it has an 8-bit data bus where the 8086 has a 16-bit data bus.)

The functions performed by the 8259A are

- Handling eight interrupts per the instructions specified in the control register (this is basically multiplexing eight interrupts down to one that is then sent to the microprocessor)

- Vectoring an interrupt request into any one of eight memory locations in the microprocessor

- Managing interrupt priorities based upon different schemes set up in the control word

- Masking individual interrupt requests
- Reading the status of any interrupt
- Being set up to handle either level-triggered or edge-triggered interrupt requests
- Handling up to 64 interrupts by nesting PICs
- Being configured to interface with either the 8085 or 8086/8088

Figure 7-11 shows the block diagram of the 8259A.

In Figure 7-11, the data bus buffer is the chip's interface with the data bus (D7–D0).

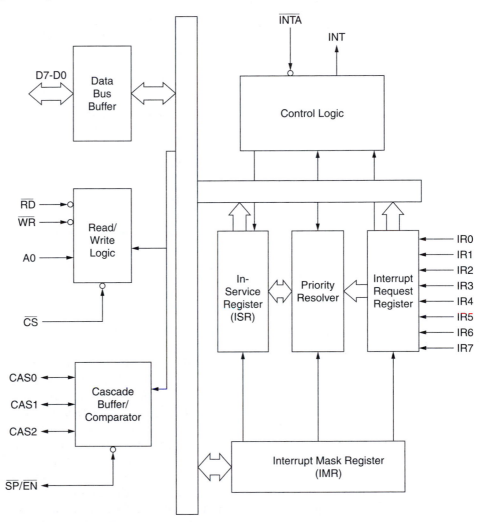

Figure 7-11 ■ An 8259A PIC block diagram

The read/write logic has the read/write input signals, the A0 input signal, and the chip select.

The cascade buffer/comparator handles the CAS0–CAS2 signals that are used when chips are cascaded. It also has the slave program/enable buffer input.

The control logic controls the operation of the chip, and sends and receives the INT and $\overline{INTA}$ signals from the microprocessor.

The interrupt request register is where the eight interrupt requests come into the PIC from the peripheral devices.

The in-service register contains the levels of interrupts that are being serviced, while the interrupt mask register contains which interrupts are being masked.

The priority resolver determines which interrupt request to handle based upon these three registers, and then notifies the control logic to send an INT signal.

Figure 7-12 shows the pinout for the 8259A PIC.

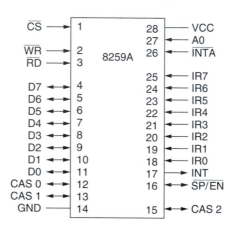

Figure 7-12 ■ An 8259A PIC pinout

Control Word

The 8259A has two control words that need to be written when initializing the chip. They are the initialization command word (ICW) and the operational command word (OCW). In order to set up the 8259A, two ICWs—ICW1 and ICW2—must be written out to the chip. Once the chip is initialized, it can be set up to operate in different modes by using three different OCWs.

Figure 7-13 shows the layout of the ICW control words.

In Figure 7-13, considering ICW1: D0 specifies whether two more ICWs are needed to initialize the 8259A. D1 specifies whether this PIC is in the single mode or the cascaded mode, while D2 specifies whether the interrupt vector addresses to be used are four apart or eight apart for each vector. D3 specifies whether the interrupts are to be edge-triggered or level-triggered, and D4 is not used. D5–D7 specify A7–A5 address bits of the interrupt vector addresses.

Control word ICW2 specifies the upper address bits—A15–A8 of the interrupt address vector. The interrupt vector addresses that the PIC sends to the 8085 are composed of A15–A5 (if the addresses are 4 apart) and the PIC provides the rest of the interrupt vector address: A15–A5 0 0 0 0 0 to A15–A5 1 1 1 0 0. If the PIC is set to addresses being 8 apart, the interrupt vector addresses are A15–A6 0 0 0 0 0 0 to A15–A6 1 1 1 0 0 0.

Modes

Many different priority modes are available in the 8259A, which can be changed during the execution of the program by using certain command words.

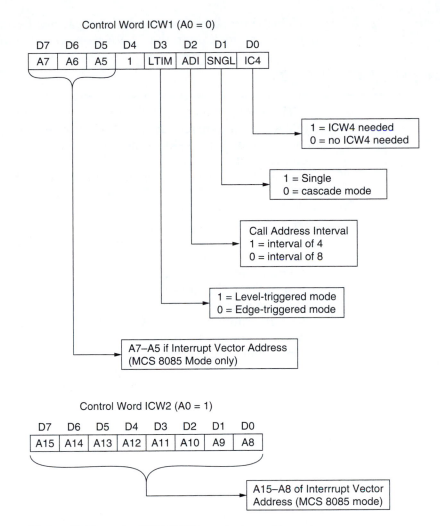

Figure 7-13 ■ An 8259A PIC control word layout

Fully Nested Mode

Fully nested mode is a simple and often used mode in which the priority of the interrupts goes from highest, IR0, to the lowest, IR7. You can also make any of the IRs the highest priority, causing the priority to go down from there. For example, IR4 can be made highest priority, giving IR5 next highest in priority, until it gets down to IR3 which would have the lowest priority.

Automation Rotation Mode

In this mode, after an interrupt is serviced, it goes to the lowest priority. Thus, the priority is a rotating scheme based on which interrupts were last serviced.

Specific Rotation Mode

In this mode, an interrupt (IR) can be specified by the program as the lowest priority, and the priorities of the rest of the interrupts are set accordingly. Therefore, it is similar to the Automation Rotation mode, except the program can specify which of the interrupts is the lowest priority.

The 8259A has other modes and features beyond what is presented here that you can look up in the 8259A manual or specification sheets.

7-4 DMAC—8237

The 8237 is the Direct Memory Access Controller (DMAC). The function of a DMAC is to transfer data at a high rate from a peripheral device (such as a floppy disk or hard disk) to memory, or vice versa. It also has a memory-to-memory mode as well.

Basically the way the DMAC works is that there are four channels in the 8237, each assigned to a specific peripheral (for example, a floppy disk or hard disk). To read data from a hard disk to memory, the microprocessor sends the start address of where the data is to be stored in memory, specifies the channel the disk is assigned to, specifies it is a read operation, and then sends the number of bytes to be moved.

The DMAC must share the system bus with the microprocessor. To do this, the DMAC requests use of the bus from the microprocessor, and when the microprocessor grants that request, the DMAC takes control of the system bus for its transfers of data. The microprocessor can do other work during this time, but cannot use the system bus. The idea of freeing up the microprocessor from handling data transfers 1 byte at a time is the main reason the DMAC was developed.

Figure 7-14 shows the block diagram of the 8237 DMAC.

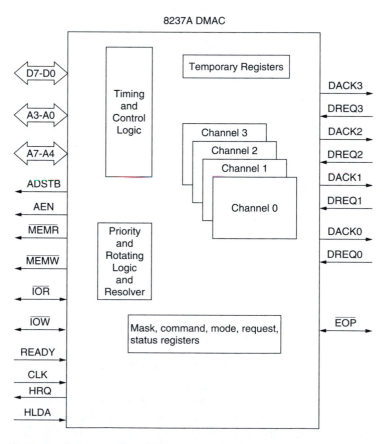

Figure 7-14 ■ An 8237 DMAC block diagram

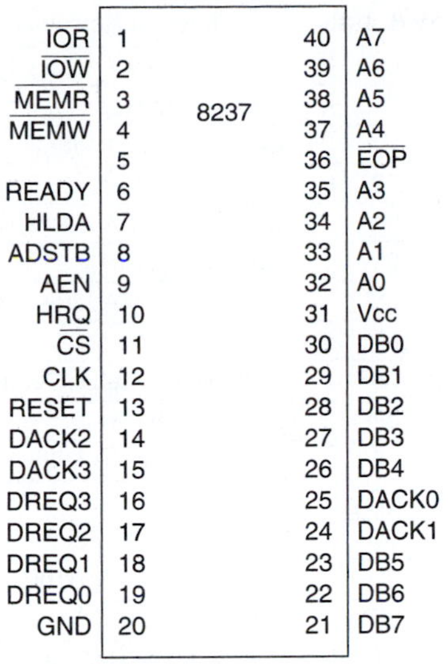

```
        IOR  | 1          40 | A7
        IOW  | 2          39 | A6
       MEMR  | 3          38 | A5
       MEMW  | 4   8237   37 | A4
             | 5          36 | EOP
      READY  | 6          35 | A3
       HLDA  | 7          34 | A2
      ADSTB  | 8          33 | A1
        AEN  | 9          32 | A0
        HRQ  | 10         31 | Vcc
         CS  | 11         30 | DB0
        CLK  | 12         29 | DB1
      RESET  | 13         28 | DB2
      DACK2  | 14         27 | DB3
      DACK3  | 15         26 | DB4
      DREQ3  | 16         25 | DACK0
      DREQ2  | 17         24 | DACK1
      DREQ1  | 18         23 | DB5
      DREQ0  | 19         22 | DB6
        GND  | 20         21 | DB7
```

Figure 7-15 ■ An 8237 DMAC pinout configuration

Figure 7-15 shows the pinout of the 8237 DMAC.

In Figure 7-14, DREQ0–DREQ3 are the input signals from peripherals such as floppy disks or hard disks. The peripherals use these inputs to the DMAC to request DMA service.

DACK0–DACK3 are the acknowledgement lines from the DMAC back to the appropriate peripheral device, telling the device its request for DMA service has been granted.

A7–A0 are address lines used by the microprocessor to specify which control register it's inputting data into and then used by the DMAC when sending out a 16-bit address.

D7–D0 represent the data bus from the microprocessor.

$\overline{MEMR}$ and $\overline{MEMW}$ are output signals from the 8237 to system memory, used when writing or reading system memory.

$\overline{IOR}$ and $\overline{IOW}$ are input and output signals used to tell the 8237 it is being written or read, as well as tell the 8237 to read or write a peripheral.

HRQ and HLDA are signals used to request (Hold Request) and get acknowledgements back (Hold Acknowledgement) when requesting the use of the system bus from the microprocessor.

AEN and ADSTB are address enable and address strobe signals. They are outputs from the DMAC that are used when generating a 16-bit address to send out on the address bus.

Each channel in the DMAC has two internal 16-bit registers: one to hold the memory address, and one to hold the count of bytes to be moved. With 16 bits for the count register, each channel can transfer no more than 64K bytes at one time.

Figure 7-16 shows the internal registers of the 8237 and the address in the DMAC of each channel.

In Figure 7-16, the registers are addressed using A3–A0 from the microprocessor. These four address lines go into the DMAC and select the appropriate register as shown. The address range of the registers displayed in the figure is 00H–0FH. This assumes that A7–A4 are all zero and that they select the $\overline{CS}$ when they are all low. A3–A0 actually go from 0000 to 1111 or 0H to FH. The second part of the address comes from the logic that has A7–A4 going into it and then into the $\overline{CS}$.

Address	Internal Register
00	Ch 0 Memory Address Reg.
01	Ch 0 Count Reg.
02	Ch 1 Memory Address Reg.
03	Ch 1 Count Reg.
04	Ch 2 Memory Address Reg.
05	Ch 2 Count Reg.
06	Ch 3 Memory Address Reg.
07	Ch 3 Count Reg.
08	Read/Write Status/Command Reg.
09	WR Request Reg.
0A	WR Single Mask Reg.
0B	WR Mode Reg.
0C	WR Clear Byte Pointer F/F
0D	R/W Master Clear/Temp. Reg.
0E	WR Clear Mask Reg.
0F	WR All Mask Reg. bits

Note: The addresses shown above assume that A7-A4 are all zero and as such activates the $\overline{CS}$ input of the 8237A. A3-A0 specify the low order part of the register address.

Figure 7-16 ■ 8237 DMAC internal registers

To program the 8237 DMAC, do the following:

1. Write a command word in the mode register that selects the channel and specifies the type of transfer—read, write, or verify—and the DMA mode (block, single byte, and so on).

2. Write a control word in the command register that specifies conditions such as the priority between channels, DREQ and DACK active levels, and timing and enables.

3. Write the start address of the data block to be transferred in the channel memory address register.

4. Write the count of the words to be transferred into the channel count register.

The actual programming of the 8237A DMAC is fairly complicated. Because of this, a manual or specification sheet of the 8237A should be consulted for more details on how to program this chip.

7-5 Mem/IO/T Chip—8155

The 8155 is a multipurpose chip that has RAM on it, three I/O ports, and a timer. Created to work specifically with the 8085, it was designed as a multipurpose chip. Thus, if you had a small microprocessor system, you would only need one support chip—the 8155—since it provides all three functions.

The RAM memory on the chip is 256 × 8 (256 bytes) static RAM. The I/O ports consist of two ports–A and B–that are 8-bit ports, and port C, which is a 6-bit port. All three ports can be configured as simple input or output ports. Ports A and B can also be programmed to use the handshaking mode in which each port uses three lines from port C for the handshaking signals. The timer is a 14-bit down counter and has four timing modes.

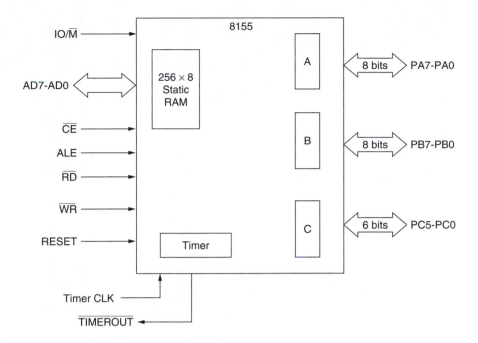

Figure 7-17 ■ An 8155 block diagram

Figure 7-17 shows a block diagram of the 8155. It also shows the following connections:

- **PA7–PA0** Port A input or output lines
- **PB7–PB0** Port B input or output lines
- **PC5–PC0** Port C input or output lines
- **Timer CLK** Timer clock inputted to the timer
- **$\overline{TIMEROUT}$** Output from the timer when count is completed
- **RESET** Resets the chip
- **$\overline{WR}$** Indicates a write into the chip registers
- **$\overline{RD}$** Indicates a read of the chip registers
- **ALE (address latch enable)** Latches the low order address AD7–AD0
- **$\overline{CE}$** Chip enable
- **AD7–AD0** Address lines from the microprocessor
- **IO/$\overline{M}$** Selects whether the RAM or the IO section of the chip is being addressed. The timer is included in the IO part.

A2	A1	A0	Port
0	0	0	Control /Status Register
0	0	1	Port A
0	1	0	Port B
0	1	1	Port C
1	0	0	LSB Timer
1	0	1	MSB Timer

Figure 7-18 shows the pinout of the 8155.

The address of the I/O ports, control register, and timer are determined by A2–A0.

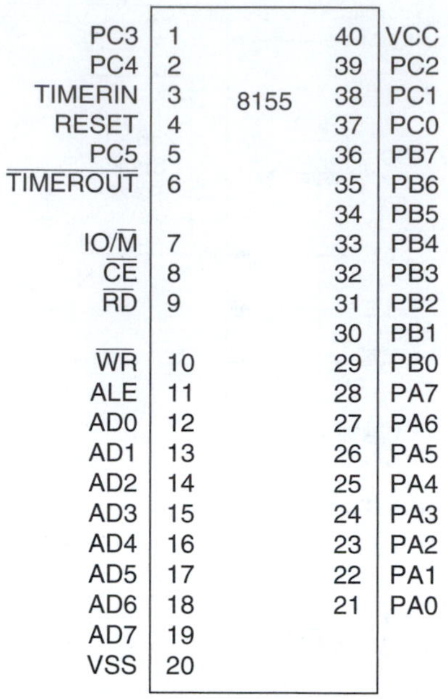

```
              PC3  | 1           40 | VCC
              PC4  | 2           39 | PC2
           TIMERIN | 3    8155   38 | PC1
             RESET | 4           37 | PC0
              PC5  | 5           36 | PB7
          TIMEROUT | 6           35 | PB6
                   |             34 | PB5
              IO/M  | 7          33 | PB4
               CE   | 8          32 | PB3
               RD   | 9          31 | PB2
                   |             30 | PB1
               WR  | 10          29 | PB0
               ALE | 11          28 | PA7
               AD0 | 12          27 | PA6
               AD1 | 13          26 | PA5
               AD2 | 14          25 | PA4
               AD3 | 15          24 | PA3
               AD4 | 16          23 | PA2
               AD5 | 17          22 | PA1
               AD6 | 18          21 | PA0
               AD7 | 19
               VSS | 20
```

Figure 7-18 ■ The 8155 pinout configuration

The rest of the address for these registers and ports shown previously are determined by the address logic that is connected to the $\overline{CE}$.

Control Word

The I/O ports and the timer are configured via a control word. Figure 7-19 shows the layout of the control word.

In Figure 7-19:

- D0 selects whether port A is input or output.
- D1 selects whether port B is input or output.
- D2–D3 specify how port C is used: all input, all output, half output and half handshaking for port A, or half handshaking for port B and half handshaking for port A.
- D4 specifies whether port A has interrupts enabled or disabled (only used in the interrupt mode).
- D5 specifies whether port B has interrupts enabled or disabled (only used in interrupt mode).
- D6–D7 are timer commands: NOP; stop counting; stop after TC; and Start.

The 8155 is a multipurpose chip that can be rather easily programmed to use the I/O ports and the timer. As shown previously, the control word is used to configure and use the I/O ports and the timer portion of the 8155. The RAM on the chip can be addressed via the address bus and the $\overline{CE}$, and does not require use of the control word.

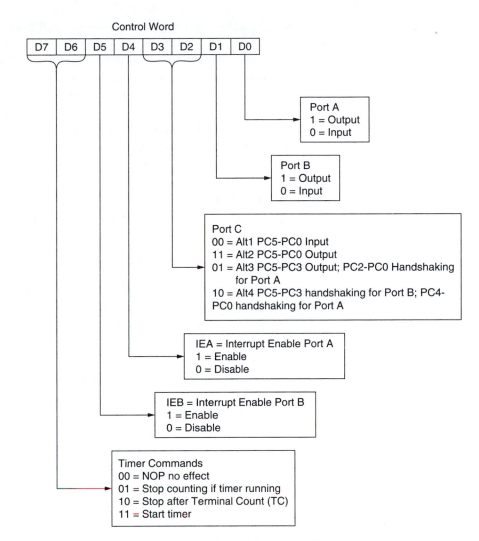

Figure 7-19 ■ The control word layout for an 8155

Summary

- Many general-purpose programmable support chips are used with a microprocessor when designing microprocessor-based systems.

- Five Intel programmable support chips are often used with the 8085 microprocessor. These are the 8255 A – Programmable Peripheral Interface (PPI), the 8254 – Programmable Interval Timer (PIT), the 8259A – Programmable Interrupt Controller (PIC), and the 8237 – Direct Memory Access Controller (DMAC).

- Programmable support chips are similar in the general way they are used. All of them have a control word that controls the mode the chip is in, as well as the configuration of the chip.

- The 8255A – Programmable Peripheral Interface (PPI) is a support chip that sets up three parallel input/output ports: ports A, B, and C.

- The 8255A has two 8-bit ports—A and B—that can be programmed as input or output, as well as port C which can be programmed as two 4-bit ports designated as either input or output.
- In the 8255A, there are two functions it can perform—the I/O function and the bit set/reset (BSR).
- The 8254 Programmable Interval Timer (PIT) is used for counting and timing. It contains three 16-bit counters that can be programmed independently into one of six different modes.
- The PIT can be used for time delays, a real-time clock, an event counter, a digital one-shot, a square-wave generator, and a complex waveform generator.
- The Programmable Interrupt Controller – 8259A manages the interrupts for the 8085, 8086, and 8088 microprocessors.
- The 8259A handles eight interrupts and can be cascaded to handle a total of 64.
- The 8237 is the Direct Memory Access Controller (DMAC) whose function is to transfer data at a high rate from a peripheral device (such as a floppy disk or hard disk) to memory, or vice versa.
- The 8237 has four channels where each channel is assigned to a specific peripheral.
- The 8155 is a multipurpose chip that has RAM on it, three I/O ports, and a timer.
- The RAM memory on the chip is 256 bytes of static RAM, 8-bit I/O ports A and B, and port C which is a 6-bit port.

Questions

1. What is the purpose of using programmable support chips?
2. What mode of configuring a programmable support chip is generally used in support chips?
3. Briefly describe the capabilities of the 8255A PPI.
4. Which port in the 8255A can be split into two 4-bit ports?
5. What are the two basic functions or modes that the 8255A can perform?
6. What happens to port C when either port A or B is in the handshaking mode?
7. What is the BSR mode in the 8255A?
8. How many timers are there in the 8254 PIT, and what size are they?
9. How does Mode 0 work in the 8254 PIT?
10. What does the 8259A PIC do functionally?
11. How does the fully-nested mode differ from the automation rotation mode in the 8259A PIC?
12. Why does using the 8237 DMAC increase overall system efficiency?
13. Explain briefly how the 8237 DMAC transfers data.
14. Why was the 8155 chip designed?
15. What is contained in the 8155 chip?

Problems

1. Write a program that sets up an 8255A PPI where port A is input, port B is output, port C upper is input, and port C lower is output. (Assume the port addressing is the same as shown in Figure 7-3.)

2. Design the addressing logic for an 8255A such that the port addresses are port A – A0H; port B – A1H; port C A2H; and the control word is A3H. Also, show the logic for the $\overline{RD}$ and $\overline{WR}$ connections.

3. Write a program to use an 8255A PPI in the BSR mode. When in the BSR mode, set bit 2 in port C to a 1.

4. Write a program to configure an 8254 PIT such that counter 1 is put in Mode 1, and the count is set to FF00H.

5. Write a program that sets up an 8259A PIC in a single chip mode, with interrupt vectors separated by eight addresses, and where the interrupt vector addresses start at 3FE0H.

6. Design the address logic such that an 8237 DMAC chip has the register addresses of 20H–2FH.

7. Write a program that sets up an 8155 so that port A is input, port B is output, and port C is input. (Assume it is not in the interrupt mode and the timer is not being used.)

Laboratory Experiments

Assumptions to Make with Regards to the Following Labs:

- The labs are to be run on an 8085 microprocessor trainer.
- The trainer should have keyboard entry and an LED display.
- The trainer has 2000H as the starting address for user RAM. If your trainer has a different starting address for users, just substitute that base address for 2000H.
- Users should be able to input hexadecimal values to enter a program.
- Hand-assembling is assumed, but if your trainer is connected to a PC and a cross-assembler is available, make the appropriate wording changes in the labs to accommodate this.
- The student can view the contents of memory and registers on the LED display on the trainer. If, however, the student is using a PC interface, make appropriate changes to recognize that the students will be viewing the memory and register contents on the PC.
- The trainer is assumed to have a single-step capability.
- Lab 10 assumes the availability of an 8255 PPI chip and the ability to connect it up to the 8085 trainer.
- Lab 11 assumes a basic logic analyzer is available, such as the Tektronix 318/338 Logic Analyzer.
- Lab 12 assumes the availability of a basic PC-based 8085 microprocessor simulator.
- The use of a microprocessor programming sheet and a tracing sheet are assumed. These forms are used when writing a program and

when tracing register and memory contents. A sample programming sheet and tracing sheet are attached at the end of this section.

This set of labs is intended to cover a complete 16-week semester where there is one three-hour lab session per week. There are 12 labs, and Labs 9 and 10 are two-week labs, so that covers 14 weeks. Usually, the first lab session is not used, and the last lab session is used for make up.

Lab 9: I/O Interfacing Lab

Introduction: This lab has the student design an I/O interfacing circuit and connect it to the 8085 microprocessor using DIP switches and Leeds.
NOTE: This lab is intended to be done over two lab sessions.
Objectives: Upon completion of this lab, the student will:

- Be able to design an interface circuit between an 8085 and switches for input
- Be able to design an interface circuit between an 8085 and Leeds for output
- Be familiar with writing a program that executes a continual loop
- Improve their troubleshooting techniques

Equipment Needed:

- An 8085 microprocessor trainer
- A prototype board
- A 74LS373 – octal transparent latch
- A 74LS244 – octal buffer
- A 74LS138 – 3-to-8 decoder
- A 74LS02 – quad two-input NOR gate
- A 40-pin IC clip
- Eight DIP switches
- Eight Leeds

Procedure:

1. It may be helpful to review the I/O interfacing circuits in the text in Chapter 4, section 4.3 "I/O Interfacing." These are similar to the circuits the student will be designing.
2. The overall design is a circuit that takes its input from eight DIP switches and then displays the data on eight LEDs. The program will be a continuous loop that reads in the eight switches and then outputs that data to the eight LEDs, afterward repeating the process.
3. First, design the input circuit using the eight DIP switches that come in through a 74LS244 octal buffer. The buffer output goes to the data bus. (The switches should be wired so that there is +5V on them through a resistor when open—and when closed, they should close to ground.)
4. Design the output circuit using eight LEDs that are driven by the 74LS373 latch. The inputs to the 74LS373 are from the data bus.
5. Design the decoding circuit which will use a 74LS138 3-to-8 decoder and two input NOR gates. The input switches should have a port address of 06 and the LEDs should have a port address of 07.

6. Write the program that reads in the input switches on port 06 and then outputs the data on port 07 to the LEDs. It should be a continuous loop.

7. **HINT:** Troubleshooting is easier if you first do just the input circuit and the decoding circuit and then test it with a small diagnostic program—or a program that just has a few lines of code that lets you see if the switch data is getting into the microprocessor. Then design the output circuit and test it separately with a small diagnostic program to see if data from the microprocessor can get out to the LEDs. Finally, when both input and output circuits are working, load the final program that reads in the input and displays it on the LEDs.

   ANOTHER HINT: Problems most often occur in circuits like this in the decoding logic, as well as in the proper enabling and gating of the input and output chips.

8. Demonstrate to your instructor the proper operation of your circuit.

Questions:

1. What did you learn about troubleshooting techniques in this lab?
2. Discuss the actual problems you had in building and testing this circuit.
3. Write a conclusion for this lab.

Lab 10: I/O Interfacing with the 8255 PPI Lab

Introduction: This lab has the student design an I/O interface to an 8085 microprocessor using an 8255 PPI (Programmable Peripheral Interface), with DIP switches and seven-segment displays.

NOTE: This lab is intended to be done over two lab sessions.

Objectives: Upon completion of this lab, the student will

- Be able to design an I/O interface through a PPI to read in DIP switches
- Learn how to write a program to setup a PPI
- Be able to design an I/O interface through a PPI to output data to seven-segment LED displays
- Learn to write a program that will convert input data to decimal for outputting

Equipment Needed:

- An 8085 microprocessor trainer
- A prototype board
- An 8255 PPI
- Four DIP switches
- Two seven-segment displays
- Two 7447 BCD seven-segment drivers
- A 74LS373 – octal transparent latch
- A 40-pin IC clip

Procedure:

1. It may prove helpful to review section 7–1 in the textbook covering the PPI, in order to brush up on how to connect an 8255, how to enable and address the PPI, and how to program one.

2. The complete circuit is an 8085 microprocessor that interfaces to a PPI, and then, through the PPI, interfaces to four DIP switches as input. A program converts the 4 bits of inputted data to two decimal digits. These values are then outputted through the PPI to two seven-segment displays.

3. The port assignments in the PPI are

   Port A – output to the most significant seven-segment LED
   Port B – output to the least significant seven-segment LED
   Port C (lower) – input from the four DIP switches

   The port addresses are

   Port A – E0H
   Port B – E1H
   Port C – E2H
   Control Word – E3H

4. First, design the input circuit that has four DIP switches that input to the PPI—the lower part of Port C. Remember, the switches should be wired so that there is +5V through a resistor when the switch is open, and ground when the switch is closed. Also, since you are going into the PPI, the switches can be wired directly into the PPI without using a buffer.

5. Design the output circuit (which has Port A going to the most significant seven-segment LED) through both a 74LS373 latch and then to a 7447 BCD seven-segment display driver, and then to the seven-segment LED. Repeat this for Port B to the least significant seven-segment LED through a 74LS373 and a 7447.

6. Design the address decoding circuit for the PPI as well as for the 74LS373. Use whatever chips you need—ANDs, ORs, NANs, NORs, and/or inverters. Remember, the port addresses for the PPI are E0H–E3H.

7. Write the program that reads in the four switches, and then converts the binary value of the inputted data to two decimal digits. (Hint: You may want to use the DAA instruction.) Then, the program outputs the least significant digit out, as well as the most significant digit out to the two seven-segment displays. The program should be a continuous loop and keep repeating the preceding process.

8. Remember, when troubleshooting your circuit, test just the input circuit, then just the output circuit before testing them together. You may also need to write small diagnostic programs to test each side of the circuit.

9. Demonstrate the proper operation of your circuit to your instructor.

Questions:

1. What did you learn about troubleshooting techniques in this lab?

2. Discuss the actual problems you had in building and testing this circuit.

3. Write a conclusion for this lab.

8

System Applications

Objectives:

Upon completion of this chapter, you should:

- Understand how multiplexed displays work, and their advantages

- Be able to interface multiplexed displays to an 8085 microprocessor

- Know how a matrix keyboard works

- Comprehend how to interface to a matrix keyboard from an 8085

- Understand interfacing to an A/D converter

- Know how to interface to a D/A converter

- Comprehend what comprises a single-board computer

Key Terms:

- **Multiplexed display**—A circuit where one output port is connected to several displays (such as seven-segment LED displays), and another output port is used to select which of the displays the data on the bus is intended for

- **Matrix keyboard**—An input device where—for example—16 keys are configured as a 4×4 matrix with four rows and four columns

- **Analog to digital (A/D) converter**—A device or circuit that takes an analog input and converts it to a digital value

- **Digital to analog (D/A) converter**—A device or circuit that converts a digital input to an analog output

- **Single-board computer**—A microprocessor system that is housed completely on a single printed circuit board

Introduction

When considering microprocessor system applications, it's important to keep certain functions or interfaces in mind since they often occur in system processes. Some of these include interfacing to a **multiplexed display**, to a **matrix keyboard**, to an **A/D converter**, to a **D/A converter**, and interfacing to a **single-board computer.**

8-1 Interfacing Multiplexed Displays

The use of **multiplexed displays** is common in microprocessor-based systems. Basically, it's done in order to reduce the extra wiring and ports needed. One output port is connected to several displays (such as seven-segment LED displays), while another output port is used to select which of the displays the data on the bus is intended for. In this way, two ports can handle multiple displays.

Figure 8-1 shows a typical multiplexed display configuration.

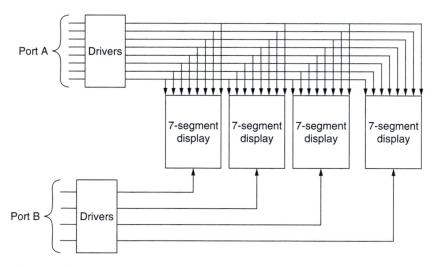

Figure 8-1 ■ A multiplexed display configuration

In Figure 8-1, port A and B could be from an 8255A PPI, for example, or from an 8155 M/IO/T, as shown in the previous chapter. Port A sends data to all of the four seven-segment displays, while the lines from port B select which display chip is enabled.

In Figure 8-1, drivers are shown between ports A and B and the LED displays. This is because the 8255A PPI and the 8155 do not output enough current for most seven-segment displays; therefore, transistor or IC drivers are needed to supply a larger current to the displays. These drivers are often called segment drivers (such as in SN75491) or digit drivers (as in the case of SN75492).

A program alternates sending the data to be displayed to each of the seven-segment displays while it selects the appropriate display via the proper line in port B. When done this way, the cycle time is fast enough that the displays appear to be continuously on and steady; however, in reality the displays are being turned on when being written and then turned off when other displays are being written. Typically, with seven-segment displays, the port B lines are connected to the common anode or cathode, depending on which type of seven-segment display chip is used.

If you did not multiplex the displays, the system would need one port per display, thus requiring four output ports for the circuit in Figure 8-1 instead of two. It should be pointed out that most chips such as the 8255A or the 8155 only have three output ports. So, more hardware and wiring would be required if the display was not multiplexed.

Figure 8-2 shows the flow chart for a routine that would display information on the circuit shown in Figure 8-1.

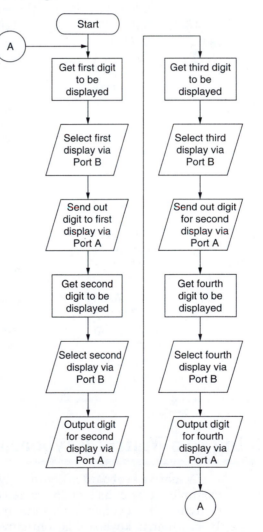

Figure 8-2 ■ A multiplex display flow chart

Figure 8-2 illustrates the routine that continuously cycles through the four displays, alternately sending out a digit to each display one at a time.

Example 8.1
▼

Problem: Write a subroutine that continually sends out four different values stored at memory locations 2020H–2023H to four multiplexed seven-segment displays. Assume that the I/O address of port A in the PPI is B0H, the I/O address of port B is B1H, and the I/O address of the control word is B3H. The lines in port B are PB0 – first digit, PB1 – second digit, PB2 – third digit, and PB3 – fourth digit. (Figure 8-1 is the assumed hardware configuration. Also, presume that the LED displays are common anode.)

Solution: The following routine is based on the flow chart in Figure 8-2.

```
        MVI C, FFH          ; initialize counter
        MVI A, 89H          ; initialize the PPI to: I/O mode, mode 0, ports A and
                            ; B output, port C input (not used)
        OUT B3H             ; send control word to PPI to initialize it
LOOP:   MVI A, 01H          ; load A to turn on first LED
        OUT B1H             ; send out signal to turn on first LED
        LDA 2020H           ; get first digit to display
        OUT B0H             ; send out first digit to first LED
        CALL DELAY          ; call a delay routine for 2 ms
        MVI A, 02H          ; load A to turn on second LED
        OUT B1H             ; turn on second LED
        LDA 2021H           ; get second digit
        OUT B0H             ; send out second digit to second LED
        CALL DELAY          ; delay 2 ms
        MVI A, 04H          ; load A to turn on third LED
        OUT B1H             ; turn on third LED
        LDA 2022H           ; get third digit
        OUT B0H             ; send out third digit to third LED
        CALL DELAY          ; delay 2 ms
        MVI A, 08H          ; load A to turn on fourth LED
        OUT B1H             ; turn on fourth LED
        LDA 2023H           ; get fourth digit
        OUT B0H             ; send out fourth digit to fourth LED
        CALL DELAY          ; delay 2 ms
        DCR C               ; decrement counter
        JNZ LOOP            ; loop back if counter is not 0
        RET                 ; return
```

▲

8-2 Interfacing to a Matrix Keyboard

A **matrix keyboard** is commonly used when several inputs are needed. For example, if you had eight or fewer inputs, you would probably just use switches. However, for additional inputs, a matrix keyboard should be used. A 16-key matrix keyboard is configured as a 4×4 matrix with four rows and four columns. Figure 8-3 shows a typical 16-key 4×4 matrix keyboard.

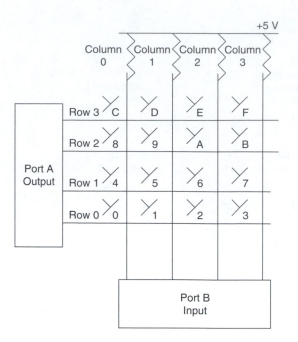

Figure 8-3 ■ A matrix keyboard

In Figure 8-3, the rows 0–3 are connected to an output port, port A as shown from a PPI. The columns 0–3 are connected to an input port, shown as port B. When none of the switches shown are pushed, none of the rows or columns are connected. When the input port B is read, all four columns will be high. (Due to the +5V connected to all four columns.) When a switch is pushed, it connects that row and that column. If a zero (low) is sent out on the row the pushed switch is on, then the column for that switch goes low. This is how a pushed switch is detected. The program must alternately send out a zero in a field of 1's on the rows, and then read the columns, looking for a low. When a low is detected, the row that was sent the zero is known, and with the row number and the column number, the program can determine which of the 16 keys are depressed.

Example 8.2
▼

Problem: Write a subroutine that interfaces with a 16-key 4×4 matrix keyboard. Assume that the rows are connected to port A from a PPI, and the columns are connected to port B from the PPI. Assume the addresses of the PPI are C0H – port A, C1H – port B, and C3H – control word. Also assume the hardware is the same as that shown in Figure 8-3, and rows 0–3 = PA0–3, and columns 0–3 = PB0–3.

Solution: First, a flow chart needs to be developed for this subroutine. Figure 8-4 shows that flow chart.

The code to implement the flow chart in Figure 8-4 is

```
MVI A, 8BH          ; set up control word for PPI – Port A output, I/O
                    ; mode, mode 0, port B input, port C input (not used)
OUT C3H             ; send control word to PPI
MVI D, 00H          ; initialize key count to 0
```

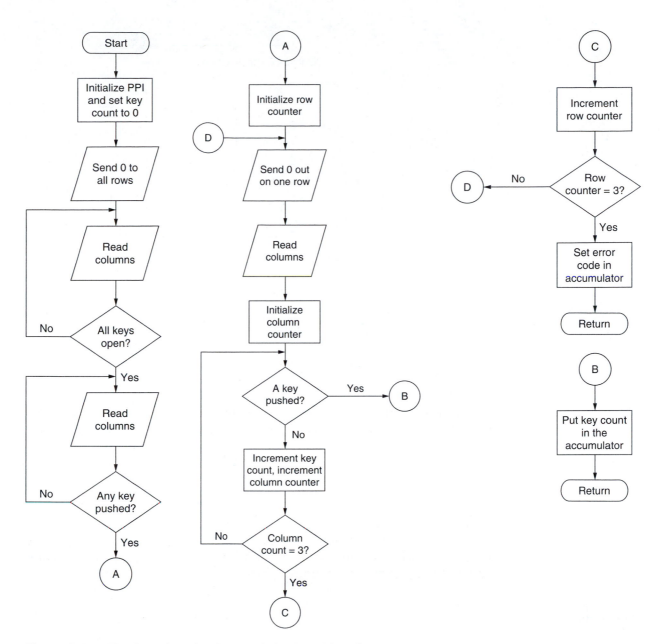

Figure 8-4 ■ The flow chart for the matrix keyboard interface

```
            MVI 00H          ; set up to send 0's out to rows
            OUT C0H          ; send out 0's to all rows
CKKYS:      IN C1H           ; read in columns
            CMA              ; complement data
            ANI 0FH          ; mask out all but PB0-3
            JNZ CKKYS        ; go back if any key was still set
NOKY:       IN C1H           ; read in columns
            CMA              ; complement data
            ANI 0FH          ; mask out all but PB0-3
            JZ NOKY          ; loop back if no key set
```

```
                MVI B,00H              ; initialize row counter
                MVI A,7FH              ; set up zero in bit 7
ROWCK:  RLC                           ; rotate 0 bit
                MOV E,A                ; save row bit location being zeroed
                OUT C0H                ; send out one zero in field of 1's
                IN C1H                 ; read in columns
                ANI 0FH                ; mask out all but PB0-3
                MVI C,00H              ; initialize column counter
COLCK:  RAR                           ; move D0 into CY
                JNC FND                ; jump if CY is zero - means found key set
                INC D                  ; increment key count
                INC C                  ; increment column count
                CPI 03H                ; see if column count =3
                JNZ COLCK              ; jump if not 3 yet
                MOV A,E                ; put row bit location back in A
                INC B                  ; increment row counter
                CPI 03H                ; see if row count =3
                JNZ ROWCK              ; jump if row not 3 yet
                MVI FFH                ; load error code of FFH into A
                RET                    ; return
FND:    MOV A,D                        ; put key count into A
                RET                    ; return
```

▲

Example 8.2 shows a typical subroutine to read a matrix keyboard and return the key that was pushed in the accumulator. It returns an FF in the accumulator for the error condition that no key was pushed after the routine sensed a key was pushed. Note that the routine begins with a loop that senses when no key is pushed before it starts looking for a pushed key. This is to make sure that the subroutine begins with a new key push.

8-3 Interfacing to an A/D Converter

Often, it is necessary for a microprocessor to interface to an **analog to digital (A/D) converter.** An A/D converter takes an analog input and converts it to a digital value, placing it on the data bus of the microprocessor. In industrial applications, it is frequently necessary to take an analog signal—for example, from an analog temperature sensor—and convert it to a digital value, afterward inputting it to the microprocessor for analysis and display.

Typically, A/D converters have an interface to the data bus of the microprocessor, an analog signal input, a reference voltage input, a start signal (to start the A/D conversion), and a ready signal to tell the microprocessor when the A/D conversion process is over and the digital output is ready to be read.

A/D converters are contained on a single chip, such as the National Semiconductor ADC0801, an 8-bit A/D converter.

Figure 8-5 shows a block diagram of an ADC0801 and how it would be connected to an 8085 microprocessor.

In Figure 8-5, the ADC0801 has an interface to the data bus of the 8085, on DB7–DB0. The 8085 has a $\overline{CS}$ input for the microprocessor to select this chip for reading or writing. The $\overline{WR}$ input on the ADC0801 serves as the start signal to begin the A/D conversion process. The $\overline{INTR}$ signal is the ADC0801's ready

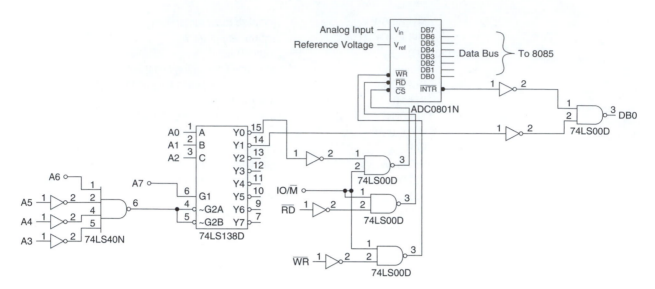

Figure 8-5 ■ An A/D converter block diagram

signal. When the ADC0801 has completed the conversion process and is ready to put the value out on the data bus, it brings the $\overline{INTR}$ signal low. The data is then put on the data bus. When the 8085 reads the data on the bus, it signals the ADC0801 by setting the $\overline{RD}$ input signal low, which resets the $\overline{INTR}$ signal.

The ADC0801 has Ain for the analog signal input, and has a Vref for inputting a reference voltage.

Example 8.3

Problem: Write a routine to read in the digital value from an A/D Converter. Assume the A/D converter is an ADC0801, as shown in Figure 8-5. (The I/O address for the ADC chip's $\overline{CS}$ is C0H, and the I/O address for reading the $\overline{INTR}$ signal is C1H, as shown in Figure 8-5.)

Solution: First, a flow chart needs to be generated for this routine. Figure 8-6 shows the flow chart to use.

The assembly language routine to implement the flow chart in Figure 8-6 is

```
          OUT C0H          ; send out an I/O Write to start conversion
LOOP:     IN C1H           ; check conversion done indicator (INTR)
          RAR              ; rotate D0 into CY
          JC LOOP          ; if D0 is 1, conversion is not yet done
          IN C0H           ; read in the data on the data bus
          RET              ; return
```

Example 8.3 shows a typical interface to an A/D converter. In this example, the conversion done signal ($\overline{INTR}$ for the ADC0801) is read in by the program. One could also implement this interface by using one of the interrupt inputs on the 8085 (for example, RST5.5) for the $\overline{INTR}$ signal to be received by. In this fashion, the program would not have to check constantly for the conversion done signal. Nevertheless, the 8085 would be interrupted when it came in. One would, however, have to write the interrupt service routine to handle this interrupt and interface back to the ADC interfacing routine.

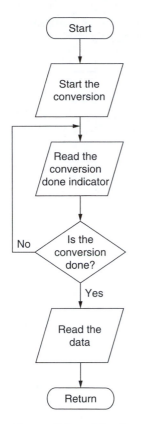

Figure 8-6 ■ The flow chart for the A/D converter example

8-4 Interfacing to a D/A Converter

Almost as common as interfacing to an A/D converter is interfacing to a D/A converter. A D/A converter does the reverse: It takes a digital output from the microprocessor and converts it to an analog signal. A D/A converter usually gives an analog output current proportional to the digital value in; thus, this output current is often sent to an op-amp to get an analog voltage proportional to the digital value in.

A D/A converter has no start signal like an A/D converter, and as soon as the data appears on the parallel data inputs, the D/A converter transforms the digital value to a proportional output analog signal. So, when interfacing with a basic D/A converter, you need a latch circuit to present and hold the digital value to the D/A converter, which then converts that value to an output current. In addition, an op-amp or similar amplifier is often connected to the chip's analog output to convert the varying output current to a varying output voltage.

However, some D/A converters are microprocessor-compatible (such as Analog Device's AD558), which means they have the latch circuits on the data bus inputs built into the chip, as well as the op-amp, in order to convert to an output voltage that's also built into the chip.

Figure 8-7 shows a block diagram of an AD558.

As shown in Figure 8-7, there are latches to hold the data coming in on the data bus, and there are two control signals: $\overline{CE}$ and $\overline{CS}$. If the $\overline{CE}$ is pulled low, the data from the bus goes into the latches. When the $\overline{CS}$ goes low, the data goes from the latches into the D/A converter. When both signals are low, the data

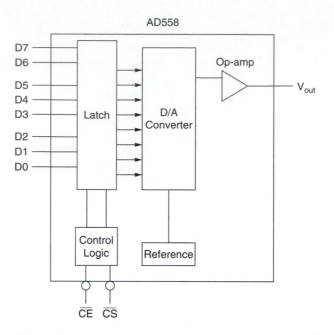

Figure 8-7 ■ The block diagram of an AD558

coming in on the bus passes directly through the latches to the D/A converter. The op-amp on the chip generates a varying analog voltage out.

Figure 8-8 shows an AD558 A/D converter interfaced to an 8085.

In Figure 8-8, the AD588 is connected to the data bus from the 8085 (the address bus (A7–A0)), while the control signals IO/$\overline{M}$ and $\overline{WR}$ are used to select the chip. The address of AFH is shown in the address logic to enable the chip. The output shown is an analog voltage out.

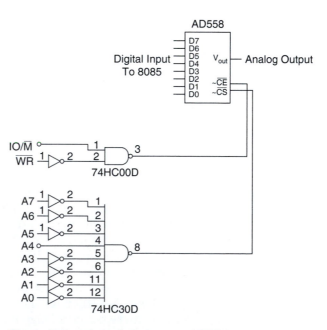

Figure 8-8 ■ Interfacing to an AD558

Example 8.4

Problem: Write a subroutine that outputs a block of memory locations (2000H–200AH) to a D/A converter and then returns. The subroutine should have a delay of 2 ms between outputs.

Solution: The flow chart for this routine is shown in Figure 8-9.

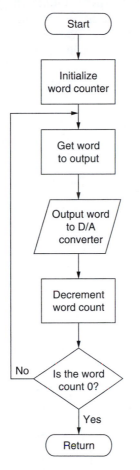

Figure 8-9 ■ Example 8.4—flow chart

The code for the flow chart from Figure 8-9 is shown next.

```
        MVI B, 0BH          ; initialize word counter to 11D
        LXI D, 2000H        ; set start memory address in DE
LOOP:   LDAX D              ; load contents of location DE points into A
        OUT AFH             ; output word to D/A converter
        CALL DELAY          ; call a 2-ms delay routine
        INX D               ; increment the memory address
        DCR B               ; decrement word counter
        JNZ LOOP            ; loop back if word count not 0
        RET                 ; return
```

Example 8.4 shows a typical routine to interface to a D/A converter. This will vary based upon the type of A/D converter used.

8-5 Single-Board Computer

A single-board computer is a microprocessor system that is totally contained on a single printed circuit board. One of the most common uses of a single-board microprocessor computer is a microprocessor trainer such as the SDK-85. These trainers are used for learning microprocessors, and contain a complete system on one board. Usually they include the microprocessor, clock, RAM, ROM, I/O ports, bus expansion drivers (to allow connections to the bus), keyboard, and seven-segment displays. Trainer boards may differ, but generally they have all of these components (or most of them) since they function as standalone single-board computers.

Figure 8-10 shows a block diagram of an SDK-85 microprocessor trainer single-board computer.

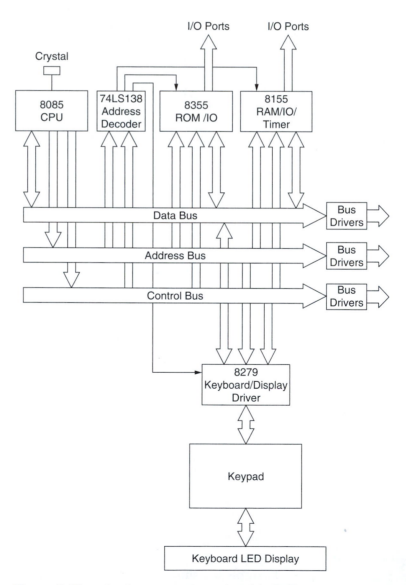

Figure 8-10 ■ A microprocessor trainer block diagram

In Figure 8-10, the 8085 is connected to the three buses. The address decoder (74LS138) shown selects the different peripheral chips, the 8355 provides ROM and some I/O ports, and the 8155 provides RAM, timers, and three I/O ports (see the previous chapter). The 8279 is a keyboard/display driver that handles the keypad and the LED display that is on the single-board computer. The bus drivers shown provide a usable interface to all three buses for connecting to external circuits.

Another common use of a single-board computer is a microprocessor industrial control system that is either rack-mounted or in some kind of ruggedized enclosure. These industrial single-board computers typically have a complete computer system on a single board—similar to the microprocessor trainer. However, the industrial control computers do not have the keypads, the seven-segment displays, and the bus expansions since those are features needed for a trainer, not an embedded controller system.

Figure 8-11 shows a block diagram of a typical industrial control single-board computer.

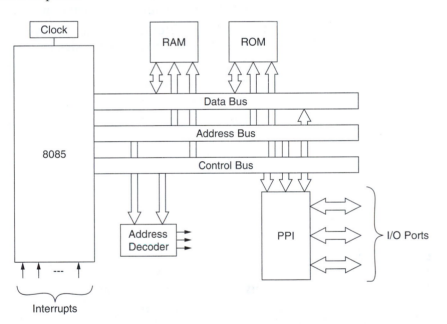

Figure 8-11 ■ An industrial control computer block diagram

As shown in Figure 8-11, the generic Industrial control computer has a microprocessor (8085), onboard RAM, onboard ROM, a PPI to provide I/O ports coming off of the board, and an address decoder used to address the onboard chips. Also, the microprocessor's interrupts are usually given off-the-board access for use by specific applications.

Summary

- The use of multiplexed displays is common in microprocessor-based systems.
- Multiplexed displays are used in order to reduce the extra wiring and ports needed.

- An interfacing routine for a multiplexed display continuously cycles through all of the displays, alternately sending out a digit to each display one at a time.

- A matrix keyboard is a commonly used input device when more than a few inputs are needed.

- A 16-key matrix keyboard is configured as a 4×4 matrix with four rows and four columns.

- It's often necessary for a microprocessor to interface to an analog to digital (A/D) converter. An A/D converter takes an analog input and converts it to a digital value, placing it on the data bus of the microprocessor.

- Typically, A/D converters have an interface to the data bus of the microprocessor, an analog signal input, a reference voltage input, and a start signal (to start the A/D conversion), and a ready signal to tell the microprocessor when the A/D conversion process is over and the digital output is ready to be read.

- A D/A converter takes a digital output from the microprocessor and converts it to an analog signal.

- Microprocessor-compatible D/A converters have latch circuits on the data bus inputs built into the chip, as well as an op-amp to convert to an output voltage, which is also built into the chip.

- A single-board computer is a microprocessor system that is totally contained on a single printed circuit board.

- A common use of a single-board computer is as a microprocessor industrial control system.

- Another common use of a single-board computer is as a microprocessor trainer.

Questions

1. Why are multiplexed displays used?
2. Explain briefly how to interface to a multiplexed display.
3. Why are segment and digit drivers needed when interfacing to LED displays?
4. What are the advantages of using a matrix keyboard over separate switches?
5. Describe how a typical A/D converter interfaces to a microprocessor.
6. Why does a typical A/D converter need a ready signal?
7. What does a microprocessor-compatible D/A converter have in it that a regular D/A converter does not?
8. What are the basic components on a single-board computer that serve as a microprocessor trainer?
9. What components does a typical single-board industrial controller have on it?
10. Name other possible uses of a single-board industrial computer.

Problems

1. Design an interface to a multiplexed seven-segment display, using an 8255 PPI and an 8085 microprocessor. Assume there are four LED displays and they are common anode. Make the PPI port addresses F0H–F3H.

2. Write a subroutine that interfaces to the multiplexed display described in Problem 1, and repeatedly sends out the values stored in memory locations 3000H–3003H.

3. Write a subroutine that interfaces to the multiplexed display described in Problem 1 and sends out the values stored in memory locations 2000H–2013H. Have the routine delay for one second between groups of four values.

4. Design an interface between an 8085, a PPI, and a matrix keyboard. The keyboard is a 20-key (five-row by four-column) matrix. Make the PPI port addresses 10H–13H.

5. Write a subroutine that interfaces to the matrix keyboard described in Problem 4. Include code at the beginning that waits until no key is pushed before it starts looking for a new key to be pushed.

6. Design an interface to an ADC0801 A/D converter and an 8085 microprocessor. Design it such that the ready signal is read by a program as opposed to coming into the 8085 as an interrupt. Assume the I/O address for the converter is 80H and 81H for the ready signal.

7. Write a subroutine to interface to the A/D converter described in Problem 6. Have the routine poll for the ready signal and return the value in the accumulator when done.

8. Design an interface to an AD558, a microprocessor-compatible D/A converter, from an 8085. Assume the I/O address for the converter is 10H.

9. Write a routine that interfaces to the D/A converter described in Problem 8. Have the routine output a block of words using memory from 2040H–2060H. The routine should delay for 4 ms between outputs.

10. At the block diagram level, design a single-board computer that has up to 16 input/outputs, has the interrupts from the 8085 accessible, and has a keypad on board for inputting. (Show the bus widths.)

Laboratory Experiments

Assumptions to Make with Regards to the Following Labs:

- The labs are to be run on an 8085 microprocessor trainer.
- The trainer should have keyboard entry and LED display.
- The trainer has 2000H as the starting address for user RAM. If your trainer has a different starting address for users, just substitute that base address for 2000H.
- Users should be able to enter hexadecimal values to enter a program.

- Hand-assembling is assumed, but if your trainer is connected to a PC and a cross-assembler is available, make the appropriate wording changes in the labs to accommodate this.

- The student can view the contents of memory and registers on the LED display of the trainer. If, however, the student is using a PC interface, make appropriate changes to recognize that the students will be viewing the memory and register contents on the PC.

- The trainer is assumed to have a single-step capability.

- Lab 10 assumes the availability of an 8255 PPI chip and the ability to connect it up to the 8085 trainer.

- Lab 11 assumes a basic logic analyzer is available such as the Tektronix 318/338 Logic Analyzer.

- Lab 12 assumes the availability of a basic PC-based 8085 Microprocessor Simulator.

- The use of a microprocessor programming sheet and a tracing sheet are assumed. These forms are used when writing a program and when tracing register and memory contents. A sample programming sheet and tracing sheet are attached at the end of this section.

This set of labs is intended to cover a complete 16-week semester where there is one three-hour lab session per week. There are 12 labs, and Labs 9 and 10 are two-week labs, so that covers 14 weeks. Usually, the first lab session is not used, and the last lab session is used for make up.

Lab 11: Logic Analyzer Lab

Introduction: This lab demonstrates the use of a logic analyzer and how it displays the data on buses when connected to a microprocessor system.

Objectives: Upon completion of this lab, students will

- Gain a basic understanding of how a logic analyzer works

- Be able to connect up and capture data from buses in a microprocessor system

- Be able to match up the data recorded by the logic analyzer with the execution of the running program

Equipment Needed:

- An 8085 microprocessor trainer

- A logic analyzer, such as the Tektronix 318/338

- A 40-pin IC clip

Procedure:

1. Familiarize yourself with how to operate the specific logic analyzer you are using. Logic analyzers in general connect to parallel data lines for display and are triggered when a specified input occurs. When the trigger occurs, the data on the inputs is "captured" and displayed. Typically, the data on the inputs can be stored before the trigger, after the trigger, or before and after the trigger. The data can usually be displayed in timing diagrams or in hexadecimal.

2. The Tektronix 318/338 has two input channels or probes—A and B—and each one has eight data inputs. Each probe has a ground

connection, but if using both A and B, only one of the probes needs to be connected to the ground.

3. Hand-assemble the program listed next and store it in the trainer starting at location 2000H. Once loaded, start the program executing.

```
START: MVI A, 42H
       MVI B, 55H
       ADD B
       JMP START
```

4. It would be helpful at this point to review in the textbook the timing diagrams of certain instructions shown in Chapter 4 since you will be looking at the data on the address bus and data bus as the preceding program cycles over and over. You'll be viewing this data in hexadecimal form.

5. Connect probe A with A8–A15. Connect probe B with AD0–AD7. Connect either the ground of A or B to the ground.

6. Following the logic analyzer's manual, go through the setup menu on the analyzer and then the trigger menu. Set the trigger up to be 20H on channel A and 00H on channel B. This corresponds to the address 2000H coming down the address bus, which is the first address of the program that you have written and stored in the microprocessor. 2000H will occur on the address bus when that address is sent down the address bus to fetch the opcode of the first instruction.

7. With the program running, and the logic analyzer connected and initialized, press Start on the logic analyzer.

8. After a few seconds, the analyzer will indicate that the trigger has been received and the data has been stored.

9. Go to the Data display on the analyzer and change the display to hexadecimal form.

10. The data should look something like the following:

A (A8–A15)	B (AD0–AD7)	
20	00	address of opcode
20	3E	opcode (MVI B) on data bus
20	01	address of data operand
20	42	operand on data bus
20	02	address of opcode
20	06	opcode (MVI B) on data bus
20	03	address of data operand
20	55	operand on data bus
20	04	address of opcode

Each set of numbers on A and B may repeat for two or three cycles since the data is being recorded faster than a single cycle time.

11. Record the data on the logic analyzer display, and identify what the data is (as shown previously in step #10). When recording the data, don't list redundant, consecutive values.

Questions:
1. Write a conclusion for this lab.

Lab 12: Microprocessor Simulator Lab

Introduction: This lab uses a microprocessor simulator to run two 8085 assembler programs.

Objectives: Upon the completion of this lab, the student will

- Be able to enter and run a program on a microprocessor simulator
- Be familiar with how a microprocessor simulator works
- Understand how a microprocessor simulator can be used as a good diagnostic tool

Equipment Needed:

- A PC-based microprocessor simulator

Procedure:

1. Familiarize yourself with your particular 8085 microprocessor simulator. Simulators may differ in certain aspects, but basically they all do the same thing. First, specify the starting address to begin loading the program. Then, enter the code. The code is input at the mnemonic level, and the hexadecimal is generated automatically. You then execute the program. The contents of both registers and memory are visible during execution—line by line. Usually, you can run the program freely, or you can single-step the program.

2. Enter the following program into the microprocessor simulator.

   ```
   START:   LDA 2020H          ; put contents of 2020H into A
            CMA                ; complement A
            ANI 80H            ; AND A with 80H
            JNZ SET1           ; jump if not zero
            MVI B, 00H         ; load 0 in B
            JMP STORE1         ; jump to STORE1
   SET1:    MVI B, 01H         ; load 1 in B
   STORE1:  LXI H, 2021H       ; load address
            MOV M, B           ; put B into 2021H
            HLT                ; halt
   ```

3. Run the simulator with the preceding code loaded. Record the contents of the registers and memory locations that change and show at what line in the program the changes occurred.

4. Enter the following code into the microprocessor simulator.

   ```
   START:   LXI SP, 20C0H      ; set up stack pointer
            MVI C, 00H         ; set counter to 0
            MVI D, 2H          ; load count in D
   START2:  DCR D              ; decrement A
            CZ CNT1            ; go to CNT1 subroutine if D is 0
            MVI A, 03H         ; put 03H into A
            XRA C              ; XOR A with C
            JNZ START2         ; repeat if C does not equal 3
            HLT                ; halt
   CNT1:    INR C              ; increment C
            MVI D, 2H          ; reset count in D
            RET                ; return
   ```

5. Run the simulator with the preceding code loaded. Record the contents of the registers and memory locations that change, and show at what line in the program the changes occurred.

Questions:

1. Describe how your simulator works, and whether it has the basic capabilities listed in the lab.

2. Discuss how a simulator could be a useful debugging tool when designing a microprocessor system.

3. Write a brief conclusion for this lab.

9

Microcontrollers

Objectives:

Upon completion of this chapter, you should:

- Understand the difference between microprocessors and microcontrollers

- Learn the basic components of a typical microcontroller

- Become familiar with the architecture and capabilities of the 8051 microcontroller

- Learn the program model and register layout of the 8051

- Understand the instruction set of the 8051

- Be able to program the 8051

Key Terms:

- **Microcontroller**—A microprocessor, RAM, ROM, timers, I/O ports, and serial com ports—all on one chip.

- **Embedded system**—A system that is embedded in a specific product to do one specific task.

- **Single-bit instructions**—Instructions that manipulate bits in the registers, the bit-addressable section of RAM, I/O ports, and the carry flag.

Introduction

Microprocessors and **microcontrollers** are different. Microprocessors are general-purpose in nature and house the microprocessor alone on a single chip. To build a system with a microprocessor, you must add external chips with RAM, ROM, timers, and I/O ports.

Basically, a microcontroller contains a microprocessor, RAM, ROM, timers, I/O ports, and serial com ports all on one chip. As such, microcontrollers are less flexible than microprocessors since each microcontroller has a fixed amount of RAM and ROM, as well as I/O ports, on the lone chip. The advantage of microprocessors is that you can build any system you want with external chips, making it more general-purpose.

Figure 9-1 shows a block diagram of a typical general-purpose microprocessor system.

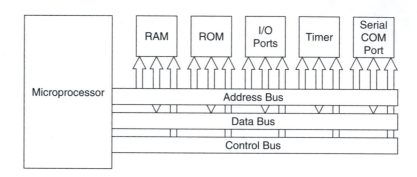

Figure 9-1 ■ The block diagram of a microprocessor system

In Figure 9-1, obviously there are several chips in this microprocessor system. Also, the mix of capabilities and functions can be varied by using different support chips.

An **embedded system** is a system that's implanted in a specific product to do one specific task. For example, a printer is an embedded system in that it has a microprocessor embedded in it that performs the particular task of interfacing the printer with a PC. An embedded system can be either a microprocessor or a microcontroller. Often, it is a microcontroller since they are more compact and economical, given they're a complete system on one chip. However, you must find a microcontroller that fits the needs of the embedded system. Other examples of embedded systems are TV remotes, microwave ovens, home security systems, copiers, telephones, answering machines, TVs, and so on.

9-1 The General Architecture of a Microcontroller

A **microcontroller** is basically a complete system on a single chip. The RAM, ROM, I/O ports, timers, and serial com port are all integrated on the chip with the microprocessor. Each microcontroller may have different amounts of each of these components, but they are still integrated on a single chip. Also, a microcontroller typically has its own instruction set. This is assembly language, and is similar to other microprocessor assembly languages, but it is nevertheless usually unique to that microcontroller.

Figure 9-2 shows the block diagram of a microcontroller.

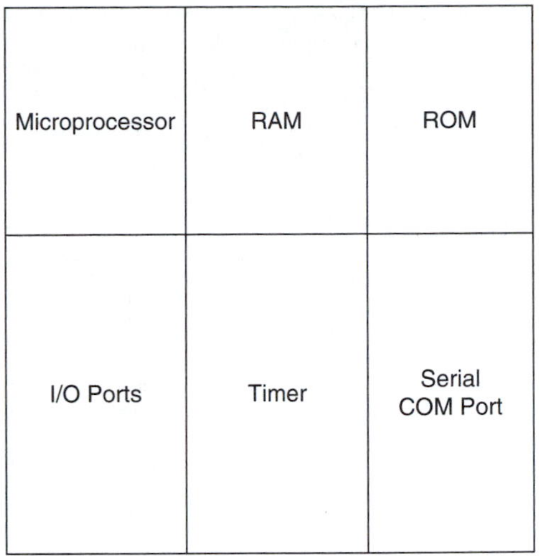

Figure 9-2 ■ The microcontroller block diagram

As in all microcontrollers, the amount of RAM, ROM, and the number of I/O ports and timers are all fixed in their quantities and capabilities. It's because these quantities are fixed that the microcontroller can be housed solely on one chip. The downside of having fixed quantities, however, is a loss of flexibility, given they cannot be changed or expanded.

But, because the microcontroller is a complete system on a single chip, they are most often the preferred choice for embedded systems, given they are more economical and take up less space.

The most common 8-bit microcontrollers are Intel's 8051, Motorola's 6811, Zilog's Z8, and the PIC 16X from Microchip Technologies. Each of these microcontrollers has a unique instruction set, and each differs in its internal register set. Therefore, a program written for one microcontroller is not compatible with any of the other microcontrollers.

Several chip manufacturers also make 16-bit and 32-bit microcontrollers. These are basically the same as the 8-bit microcontrollers in that they are a complete system on a single chip, except that they are 16 bits wide and 32 bits wide.

In selecting a microcontroller for a specific application, you should consider the following: whether the capabilities of the microcontroller meet the computing needs of the application; whether there are software tools available, such as compilers, assemblers, and debuggers; and whether the chip itself is readily available from a reliable source.

9-2 The 8051 Microcontroller Architecture

The 8051 is Intel's commonly used 8-bit microcontroller. It is a typical microcontroller in that it has all the typical components of a microcontroller on a lone chip.

Figure 9-3 shows the block diagram of an 8051.

All the components shown in Figure 9-3 lie on a single chip. The 8051 has the following:

- 4K bytes of ROM
- 128 bytes of RAM

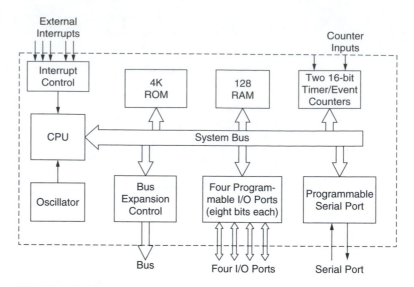

Figure 9-3 ■ The 8051 block diagram

- Two 16-bit timer/event counters
- Four 8-bit programmable timer/event counters
- One serial I/O port
- Five interrupt lines

Figure 9-3 shows all of the components from the preceding list, as well as the bus expansion capability which allows the system bus to interface if needed.

Figure 9-4 displays the pinout of the 8051.

In Figure 9-4, you can see that most of the 40 pins of the 8051 have a dual use, or are shared. The four I/O ports—P0, P1, P2, and P3—are 8-bit ports each. Dedicated pins include Vcc, RST, XTAL1 and XTAL2, GND, and $\overline{EA}$.

P1.0	1		40	Vcc
P1.1	2		39	P0.0(AD0)
P1.2	3	8051	38	P0.1(AD1)
P1.3	4		37	P0.2(AD2)
P1.4	5		36	P0.3(AD3)
P1.5	6		35	P0.4(AD4)
P1.6	7		34	P0.5(AD5)
P1.7	8		33	P0.6(AD6)
RST	9		32	P0.7(AD7)
(RXD)P3.0	10		31	$\overline{EA}$/Vpp
(TXD)P3.1	11		30	ALE/$\overline{PROG}$
($\overline{INT0}$)P3.2	12		29	$\overline{PSEN}$
($\overline{INT1}$)P3.3	13		28	P2.7(A15)
(TO)P3.4	14		27	P2.6(A14)
(T1)P3.5	15		26	P2.5(A13)
($\overline{WR}$)P3.5	16		25	P2.4(A12)
($\overline{RD}$)P3.7	17		24	P2.3(A11)
XTAL2	18		23	P2.2(A10)
XTAL1	19		22	P2.1(A9)
GND	20		21	P2.0(A8)

Figure 9-4 ■ The 8051 pinout

Vcc is the supply voltage, RST is the reset signal, XTAL1 and 2 are the crystal or clock connections, GND is the ground, and $\overline{EA}$/Vpp is the external access pin or the Vpp pin. If the ROM on the chip is being used, this pin is tied to Vcc (Vpp), but if the ROM is external, this pin is tied to the ground representing external access. Also, in conjunction with the $\overline{EA}$ pin, if an external ROM is used to store program code, the $\overline{PSEN}$ (program store enable) pin is used as an output, and goes low to activate an external ROM holding code to be executed. The $\overline{EA}$ pin must also be low to indicate that an external ROM is being used.

When the ALE/$\overline{PROG}$ pin is a 0, the 8051 uses P0.0–P0.7 as data, and when ALE goes high, the 8051 uses P0.0–P0.7 as the lower 8 bits of the address bus.

P1.0–P1.7 are the pins of the 8-bit I/O port 1. P0.0–P0.7 are port 0, and P2.0–P2.7 are port 2, while P3.0–P3.7 are port 3. Only the pins of port 1 are not shared with any other functions (as seen in Figure 9-4). Port 0 pins are shared with AD0–AD7, the lower 8 bits of the address bus as well as the data bus. When ALE is a 1, port 0 is the lower 8 bits of the address bus. When ALE is 0, port 0 is the data bus.

Port 2 pins are shared with the upper 8 bits of the address bus, A8–A15. These are used as the upper part of the address bus when ALE goes high.

Port 3 pins are shared with: TXD and RXD, or transmit and receive serial data pins; $\overline{INT0}$ and $\overline{INT1}$ for interrupts; T0 and T1 for the two timers; and $\overline{WR}$ and $\overline{RD}$ control signals for use with external memory.

So even though the 8051 is a self-contained microcontroller with RAM, ROM, timers, I/O ports, and serial ports, it still has the capability to connect to external ROM and/or RAM. Providing this additional flexibility is why so many of the pins on the 8051 have dual use.

There are other variations of the 8051 that are basically the same chip, except they have more ROM and/or more RAM. The 8052 has 8K ROM and 256 RAM, while the 8031 has zero ROM and 128 RAM.

9-3 Programming the 8051

In learning to program the 8051, you must first understand the programming model of the 8051 microcontroller. The 8051 is a fixed system in that there are always 4K bytes of ROM and 128 bytes of RAM. The address spectrums are separate for the program memory (ROM) and the data memory (RAM).

Program Memory

The 8051 has 4K bytes of ROM. This is the program memory. The address spectrum of this program memory is 0000H to 0FFFH, and is a separate address space from the data memory.

Figure 9-5 shows the layout of the program memory.

Figure 9-5 shows the address spectrum is 0000H–0FFFH in hexadecimal and 0–4095 in decimal. The ROM is 1 byte wide or 8 bits wide.

Data Memory

The data memory is RAM, and on the 8051 it is 128 bytes long and 8 bits or 1 byte wide. The address range is 00H–7FH, representing a separate address space from the program memory.

Figure 9-6 shows the layout of the data memory.

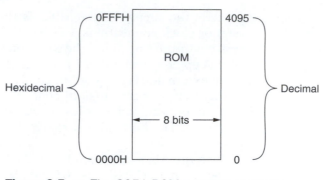

Figure 9-5 ■ The 8051 ROM program memory

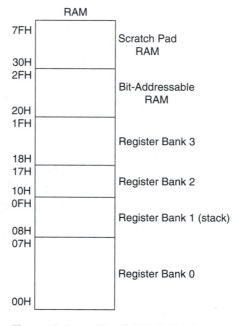

Figure 9-6 ■ The 8051 RAM data memory

Figure 9-6 shows that the RAM is divided into: register bank 0, register bank 1 (and the stack), register bank 2, register bank 3, bit-addressable RAM, and the RAM scratch pad. Remember, the entire RAM for the 8051 is only 128 bytes. 32 bytes of RAM are set aside to provide four banks of registers, each of which contains eight registers (R0–R7). Bank 1 is also shared with the stack. This can cause problems for programmers, so it's better not to use register bank 1 if you're using the stack—or don't use the stack if you want to use register bank 1. If a programmer uses the stack, the 8 bytes of bank 1 are used for the stack. The 16 bytes set aside for bit-addressable RAM are the only bit-addressable RAM in the 8051. All other RAM is byte-addressable only. A programmer can use special bit-addressable instructions and use this section of RAM (address 20H–27H) for bit-addressable operations. This capability can be very useful in a program when all you want to do, for example, is set a flag for a certain event having occurred.

The Scratch Pad section of the RAM is 80 bytes long. It is usable for any temporary storage or can be employed as a scratch pad for any calculations that a program may need to perform.

8051 Registers

The main registers used in the 8051 are A (accumulator), B, R0–R7, DPTR (data pointer), SP (stack pointer), and **PC (program counter)**. A, B, R0–R7, and SP are all 8-bit registers, while DPTR and PC are both 16-bit registers. Registers R0–R7, on the other hand, are the programmable registers, and A, B, DPTR, SP, and PC are considered special registers called special function registers (SFRs). These are discussed in the next section.

As mentioned earlier, there are banks in RAM that are usable for the programmable registers R0–R7.

Figure 9-7 shows these register banks for R0–R7.

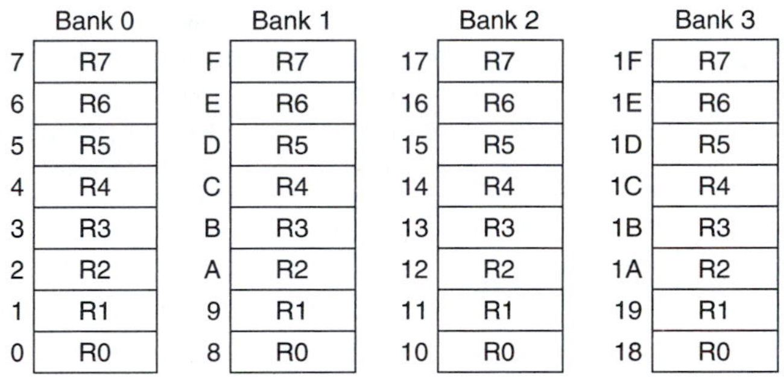

	Bank 0		Bank 1		Bank 2		Bank 3
7	R7	F	R7	17	R7	1F	R7
6	R6	E	R6	16	R6	1E	R6
5	R5	D	R5	15	R5	1D	R5
4	R4	C	R4	14	R4	1C	R4
3	R3	B	R3	13	R3	1B	R3
2	R2	A	R2	12	R2	1A	R2
1	R1	9	R1	11	R1	19	R1
0	R0	8	R0	10	R0	18	R0

Figure 9-7 ■ Register banks

Banks 0–3 are 8 bytes each, and their hexadecimal addresses are shown in Figure 9-7.

When the 8051 is initially powered up, the registers R0–R7 are automatically assigned to bank 0. In order to use one of the other banks of registers, the program must set bits D3 and D4 in the SFR register called the program status word (PSW). Table 9.1 shows how bits D3 and D4 select the bank of registers.

Bank	PSW Bit D4	PSW Bit D3
0	0	0
1	0	1
2	1	0
3	1	1

Table 9.1 ■ PSW Bits for Bank Selection

The registers can be addressed by either R0–R7, or by their absolute physical addresses, based on what bank is being used. Addressing them by R0–R7 is preferable and helps to avoid potential problems that could be caused by using the absolute addresses.

SFR Registers

Special function registers (SFRs) are normally addressed by their symbol (the preferred system), but this can also be done using an address that is outside the usable RAM address range.

Previously mentioned SFR registers include A, B, DPTR, SP, PSW, and PC. Other SFR registers exist, however, all of which are listed in Table 9.2.

Symbol	Name	Address
A*	Accumulator	0E0H
B*	B register	0F0H
PSW*	Program status word	0D0H
SP	Stack pointer	81H
DPTR	Data pointer 2 bytes	
DPL	Low byte	82H
DPH	High byte	83H
P0*	Port 0	80H
P1*	Port 1	90H
P2*	Port 2	0A0H
P3 *	Port 3	0B0H
IP*	Interrupt priority control	0B8H
IE*	Interrupt enable control	0A8H
TMOD	Timer/counter mode control	89H
TCON*	Timer/counter control	88H
T2CON*	Timer/counter 2 control	0C8H
T2MOD	Timer/counter 2 mode control	0C9H
TH0	Timer/counter 0 high byte	8CH
TL0	Timer/counter 0 low byte	8Ah
TH1	Timer/counter 1 high byte	8DH
TL1	Timer/counter 1 low byte	8BH
TH2	Timer/counter 2 high byte	0CDH
TL2	Timer/counter 2 low byte	0CCH
RCAP2H	T/C 2 capture reg high byte	0CBH
RCAP2L	T/C 2 capture reg low byte	0CAH
SCON*	Serial control	98H
SBUF	Serial data buffer	99H
PCON	Power control	87H
*Bit-addressable		

Table 9-2 ■ SFR Registers

Many SFR registers are available, but those most commonly used are A, B, PSW, SP, and DPTR. The other SFR registers are used in special cases that they apply to.

Also, note that all of the addresses of the SFR registers are beyond the RAM address range of 00H–7FH. Again, using the symbols of the SFR registers is far preferable to using the address.

The PSW (Program Status Word) Register

The PSW register is an 8-bit register and is also called the flag register. It contains four flag bits, as well as bits used for selection of the register bank as discussed earlier and shown in Table 9.1. It also has two general-purpose bits that are user-definable.

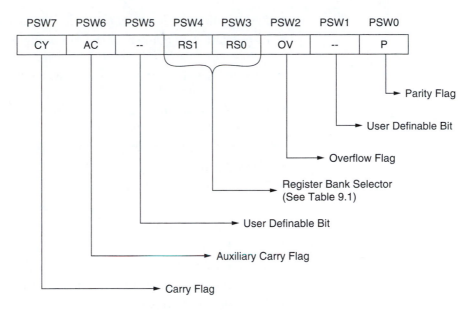

PSW7	PSW6	PSW5	PSW4	PSW3	PSW2	PSW1	PSW0
CY	AC	--	RS1	RS0	OV	--	P

Parity Flag

User Definable Bit

Overflow Flag

Register Bank Selector
(See Table 9.1)

User Definable Bit

Auxiliary Carry Flag

Carry Flag

Figure 9-8 ■ The PSW register layout

Figure 9-8 shows the layout of the PSW register.
As shown in Figure 9-8:

- **The CY** The carry flag is set whenever there is a carry bit coming out from bit D7.
- **The AC** The auxiliary carry flag is set whenever there is a carry from bits D3 to D4.
- **The P** The parity flag is set whenever there is an odd number of 1's in the accumulator.
- **The OV** The overflow flag is set whenever the result of a signed number operation is too large, causing overflow into the sign bit.

8051 Assembly Language

The 8051 assembly language instruction set can be broken down into five basic groups: data transfer, arithmetic, logical, Boolean, and branching.
The format of 8051 assembly language is the same as the 8085 in that it is

- label: opcode destination, source; comment

or

- label: opcode operand; comment

The opcode is followed by either an operand or the destination, source. Comments are preceded by a semi-colon.

Data Transfer Instructions

The data transfer instructions move or transfer data in all combinations between:

- Accumulator *(shown in the following list as A)*
- Registers R0–7 *(shown in the following list as Rn)*
- RAM(direct) *(shown in the following list as direct)*

- RAM(indirect) *(shown in the following list as @Ri)*
- Immediate data *(shown in the following list as #data for 8 bits of data and #data16 for 16 bits of data)*
- External RAM (8-bit address) *(shown in the following list as @Ri, but with a special opcode of MOVX)*
- External RAM (16-bit address) *(shown in the following list as @DPTR with MOVX opcode)*

The 8051 actual data transfer instructions are

- **MOV A,Rn** Move register to A.
- **MOV A,direct** Move direct byte to A.
- **MOV A,@Ri** Move indirect RAM to A.
- **MOV A,#data** Move immediate data to A.
- **MOV Rn,A** Move A to register.
- **MOV Rn,direct** Move direct byte to register.
- **MOV Rn,#data** Move immediate data to register.
- **MOV direct, A** Move A to direct byte.
- **MOV direct, Rn** Move register to direct byte.
- **MOV direct, direct** Move direct byte to direct byte.
- **MOV direct, @Ri** Move indirect byte to direct byte.
- **MOV direct, #data** Move immediate data to direct byte.
- **MOV @Ri, A** Move A to indirect byte.
- **MOV @Ri, direct** Move direct byte to indirect byte.
- **MOV @Ri, #data** Move immediate data to indirect byte.
- **MOV DPTR, #data16** Move 16-bit data to DPTR—data pointer.
- **MOVC A, @A +DPTR** Move code byte relative to DPTR to A.
- **MOVC A,@A+PC** Move code byte relative to PC to A.
- **MOVX A,@Ri** Move external RAM (8-bit address) to A.
- **MOVX A, @DPTR** Move external RAM (16-bit address) to A.
- **MOVX @Ri, A** Move A to external RAM (8-bit address).
- **MOVX @DPTR, A** Move A to external RAM (16-bit address).
- **PUSH direct** Push direct byte onto stack.
- **POP direct** Pop direct byte from stack.
- **XCH A, Rn** Exchange register with A.
- **XCH A, direct** Exchange direct byte with A.
- **XCH A, @Ri** Exchange indirect RAM with A.
- **XCHD A, @Ri** Exchange low-order digit indirect RAM with A.

Note: Rn = register R0–R7
Direct = 8-bit internal data location's address
@Ri = 8-bit internal data RAM location addressed indirectly through register R0 or R1
#data = 8-bit constant
#data16 = 16-bit constant
addr16 = 16-bit destination address
addr11 = 11-bit destination address (used by ACALL and SJMP)

rel = signed (2's complement) 8-bit offset byte (used by
SJMP and all conditional jumps)

bit = direct-addressed bit in internal data RAM

This is a complete list of the data transfer instructions of the 8051.

Example 9.1

▼

Problem: Write a program in 8051 assembly language that loads 00H to 07H into the registers R0–R7 and then stores them in RAM at locations 30H to 37H.

Solution:

```
MOV R0, #00H        ; store 00H in R0
MOV 30H, R0         ; put contents of R0 into RAM at location 30H
MOV R1, #01H        ; store 01H in R1
MOV 31H, R1         ; put contents of R1 into RAM at 31H
MOV R2, #02H        ; store 02H in R2
MOV 32H, R2         ; put contents of R2 into RAM at 32H
MOV R3, #03H        ; store 03H in R3
MOV 33H, R3         ; put contents of R3 into RAM at 33H
MOV R4, #04H        ; store 04H in R4
MOV 34H, R4         ; put contents of R4 into RAM at 34H
MOV R5, #05H        ; put 05H in R5
MOV 35H, R5         ; put contents of R5 into RAM at 35H
MOV R6, #06H        ; put 06H into R6
MOV 36H, R6         ; put contents of R6 into RAM at 36H
MOV R7, #07H        ; put 07H into R7
MOV 37H, R7         ; put contents of R7 into RAM at 37H
```

▲

Arithmetic Instructions

The **arithmetic operations** available in the 8051 microcontroller are

- **Add**
- **Subtract**
- **Increment**
- **Decrement**
- **Multiply**
- **Divide**
- **Decimal Adjust** Adjust the result in the accumulator when doing BCD addition.

A listing of the actual **arithmetic instructions** are shown next.

- **ADD A, Rn** Add register to A
- **ADD A, direct** Add direct byte to A
- **ADD A, @Ri** Add indirect RAM to A
- **ADD A, #data** Add immediate data to A
- **ADDC A, Rn** Add register to A with carry
- **ADDC A, direct** Add direct byte to A with carry
- **ADDC A, @Ri** Add indirect RAM to A
- **ADDC A, #data** Add immediate data to A with carry

- **SUBB A, Rn** Subtract register from A with borrow
- **SUBB A, direct** Subtract direct byte from A with borrow
- **SUBB A, @Ri** Subtract indirect RAM from A with borrow
- **SUBB A, #data** Subtract immediate data from A with borrow
- **INC A** Increment A
- **INC Rn** Increment register
- **INC direct** Increment direct byte
- **INC @Ri** Increment indirect RAM
- **DEC A** Decrement A
- **DEC Rn** Decrement register
- **DEC direct** Decrement direct byte
- **DEC @Ri** Decrement indirect RAM
- **INC DPTR** Increment data pointer
- **MUL AB** Multiply A and B
- **DIV AB** Divide A by B
- **DA A** Decimal adjust A for use with BCD addition only

This is a complete list of the arithmetic instructions available in the 8051.

Example 9.2

▼

Problem: Write a program that adds up all of the numbers stored in RAM at locations 30H and 31H, subtracts the number stored at location 32H, and then stores the result at location 33H.

Solution:

```
MOV A, #00H          ; clear A
MOV A, 30H           ; move contents of location 30H into A
ADD A, 31H           ; add contents of 31H to A
SUBB A, 32H          ; subtract contents of 32H from A
MOV 33H, A           ; put results into RAM location 33H
```

▲

Logical Instructions

The logical operations allowed in the 8051 are

- AND
- OR
- Exclusive OR
- Clear
- Complement
- Rotate
- Swap

The actual **logical instructions** are listed next.

- **ANL A, Rn** AND register to A
- **ANL: A, direct** AND direct byte to A
- **ANL A, @Ri** AND indirect RAM to A

- **ANL A, #data** AND immediate data to A
- **ANL direct, A** AND A to direct byte
- **ANL direct, #data** AND immediate data to direct byte
- **ORL A, Rn** OR register to A
- **ORL A, direct** OR direct byte to A
- **ORL A, @Ri** OR indirect RAM to A
- **ORL A, #data** OR immediate data to A
- **ORL direct, A** OR A to direct byte
- **ORL direct, #data** OR immediate data to A
- **XRL A, Rn** EX OR register to A
- **XRL A, direct** EX OR direct byte to A
- **XRL A, @Ri** EX OR indirect RAM to A
- **XRL A, #data** EX OR immediate data to A
- **XRL direct, A** EX OR A to direct byte
- **XRL direct , #data** EX OR immediate data to A
- **CLR A** Clear A
- **CPL A** Complement A
- **RL A** Rotate A left
- **RLC A** Rotate A left through carry
- **RR A** Rotate A right
- **RRC A** Rotate A right through carry
- **SWAP A** Swap nibbles in A

This is a complete list of the logical instructions available in the 8051.

Example 9.3

Problem: Write a program that reads in RAM location 3AH and checks whether it is equal to 47H. If it is, set a 1 in RAM location 3BH; if it's not, set a 0 in location 3BH.

Solution:

```
        MOV A, 3AH          ; read in contents of location 3AH
        MOV R0, #47H        ; load 47H into R0
        CPL  A              ; complement value in A
        ANL  A, R0          ; AND A and R0
        JZ FND              ; jump to FND if they were equal
        MOV 3BH, #00H       ; put 0 in location 3BH
FND:    MOV 3BH, #01H       ; put 1 in location 3BH
```

Note: The JZ or jump-on-zero instruction was used in the previous example, even though it has not been covered yet. It will be covered later in this section.

Single-Bit Instructions

Microprocessors typically can only access data by the byte. The byte is the smallest addressable group of bits whether it is RAM, I/O ports, registers, or ROM.

However, the 8051 microcontroller has the capability to address RAM, registers, I/O ports, and carry on a bit basis. (ROM only contains program instructions so there is no need to access ROM on a bit basis.) The 8051 uses its **single-bit instructions** to manipulate bits in the registers, the bit-addressable section of RAM, I/O ports, and the carry flag. This capability is very useful in many applications where a program is often needed to set one bit or read in 1 bit. In a microprocessor, the program needs to read in a byte and then AND out all but the 1 bit that is needed to be checked. Using the single-bit instructions in the 8051, however, the program can read in (or set) a single bit in a register, or in RAM or an I/O port.

A complete listing of the single-bit instructions is shown next.

- **CLR C** Clear carry
- **CLR bit** Clear direct bit
- **SETB C** Set carry
- **SETB bit** Set direct bit
- **CPL C** Complement carry
- **CPL bit** Complement direct bit
- **ANL C, bit** AND direct bit to carry
- **ANL C,/bit** AND complement of direct bit to carry
- **ORL C, bit** OR direct bit to carry
- **ORL C,/bit** OR complement of direct bit to carry
- **MOV C, bit** Move direct bit to carry
- **MOV bit, C** Move carry to direct bit
- **JC rel** Jump if carry is set
- **JNC rel** Jump of carry is not set
- **JB bit, rel** Jump if direct bit is set
- **JNB bit, rel** Jump if direct bit is not set
- **JBC bit, rel** Jump if direct bit is set and clear bit

This is a complete list of the single-bit instructions available in the 8051. The following sections show the use of single-bit instructions in the areas of I/O ports, registers, RAM, and the carry flag.

I/O Port Bit-Addressing

When addressing a single bit in an I/O port, the instruction uses the form: "SETB P1.5". This sets the 5 bit in port 1. The I/O ports in the 8051 are P0–P3. All bits in every port are bit-addressable: P0.0–P0.7, P1.0–P1.7, P2.0–P2.7, and P3.0–P3.7.

Example 9.4

▼

Problem: Read in bit 2 in Port 2, and if it is set to a 1, set bit 0 to 1 in port 3, otherwise set bit 1 to 1 in port 3.

Solution: The code to satisfy the preceding problem is shown next.

```
            JB P2.2, GOOD          ; if bit 2 in port 2 is set to a 1 jump to label GOOD
            SETB P3.1              ; set bit 1 in port 3
GOOD:       SETB P3.0             ; set bit 0 in port 3
```

The JB command jumps to the label if the bit listed is set to a 1. The SETB instruction sets the bit listed to a 1. A complete list of bit-addressable instructions is listed at the end of this section.

Register Bit-Addressing

When addressing a single bit in a register, the only registers that are bit-addressable are B, PSW, IP, IE, ACC, SCON, and TCON. These are all SFR or Special Function Registers. The form for addressing these registers on a bit-basis is "SETB B.2", which sets the second bit in the B register. Note that when bit-addressing the accumulator, the mnemonic used is ACC, not A.

Example 9.5

Problem: Write a program that checks whether bit 7 is set in the accumulator. If it is, have bit 7 set in the B register. If it's not set, then set bit 7 in the B register to a 0.

Solution: The code for this program is shown next.

```
        JB ACC.7, LABEL        ; if bit 7 in the acc is set, go to LABEL
        CLR B.7                ; sets bit 7 in B to 0
LABEL:  SETB B.7               ; sets bit 7 in B to a 1
```

The CLR instruction clears a bit or sets it to 0.

RAM Bit-Addressing

When addressing a single bit in bit-addressable RAM, the program uses the 8-bit address of the bit in the bit-addressable section of RAM.

Figure 9-9 shows the bit addresses of the bit-addressable section of RAM.

Figure 9-9 shows that every bit in the bit-addressable RAM has a unique address. The form for addressing these bits in RAM is "SETB 40H", which sets bit 40H in the bit-addressable area of RAM.

Example 9.6

Problem: Write a program that reads in the status of port 3 bit 1 and stores it in RAM location 3AH, and then reads in the status port 3 bit 2 and stores it in RAM at location 3BH.

Solution: The code for this program is shown next.

```
        JB P3.1, HIGH          ; jump if bit 1 in port 3 is set
        CLR 3AH                ; set 3AH to zero
        SJMP BITWO             ; jump to look at second bit
HIGH:   SETB 3AH               ; set 3AH to 1
BITWO:  JB P3.2, HIGH2         ; jump if bit 2 in port 3 is set
        CLR 3BH                ; set 3BH to zero
HIGH2:  SETB 3BH               ; set 3BH to 1
```

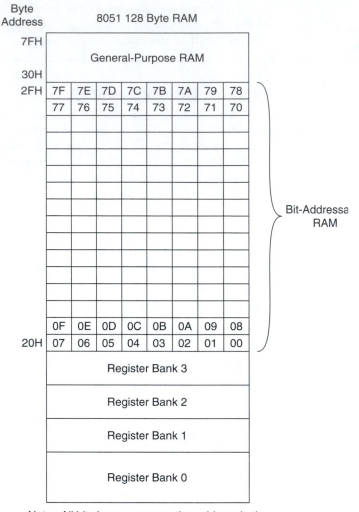

Figure 9-9 ■ Bit-addressable RAM

Single-Bit Operations with Carry

The carry flag is set and reset by arithmetic and logical instructions, but in the 8051 the carry flag may also be set and reset directly. The form of the instruction for setting or resetting the carry bit is: "SETB C". This instruction sets the carry flag to a 1. The instructions that can be used with the carry flag are SETB, CLR, CPL, MOV, JNC, JC, ANL, and ORL. These instructions are listed with all of the single-bit instructions later in this section.

Example 9.7

▼

Problem: Write a program that reads in a complete byte from port 1, and then rotates it into the carry flag, and if carry is set, sets a 1 in RAM at location

22H, and then clears the carry flag. If the carry flag is not set, have it set a 0 in RAM 22H.

Solution: The code for this program is shown next.

```
          MOV A, P1        ; read in a byte from port 1 into A
          RRC A            ; rotate A right through carry
          JC  BITON        ; jump if carry is set
          CLR 22H          ; set RAM location 22H to zero
BITON:    SETB 22H         ; set RAM location 22H to a 1
          CLR C            ; clear carry
```

Branching Instructions

All instruction sets need branching instructions or control transfer instructions. These instructions include unconditional jumps, conditional jumps, calls to subroutines, and returns from subroutines. The 8051 has also included in its instruction set a specialized "compare and conditional jump" and a "decrement and conditional jump." The branching instructions allow the use of relative addresses as well as absolute addresses.

A complete listing of branching instructions is shown next.

- **ACALL addr11** Absolute subroutine call
- **LCALL addr16** Long subroutine call
- **RET** Return from subroutine
- **RETI** Return from interrupt
- **AJMP addr11** Absolute jump
- **LJMP addr16** Long jump
- **SJMP rel** Short jump (relative address)
- **JMP @A+DPTR** Jump indirect relative to the DPTR
- **JZ rel** Jump if accumulator is zero
- **JNZ rel** Jump if accumulator is not zero
- **CJNE A, direct, rel** Compare direct byte to Acc, and jump if not equal
- **CJNE A, #data, rel** Compare immediate data to A and jump if not equal
- **CJNE Rn, #data, rel** Compare immediate data to Rn and jump if not equal
- **CJNE #Ri, #data, rel** Compare immediate data to Ri and jump if not equal
- **DJNZ Rn, del** Decrement register and jump if not equal to zero
- **DJNZ direct, rel** Decrement direct byte and jump if not zero
- **NOP** No operation

This is a compete list of branching instructions for the 8051.

Example 9.8

Problem: Write a program that zeroes out a block of ten RAM locations, starting at location 33H.

Solution: The code for the program is shown next.

```
        MOV R1, #00H          ; load word count in R1
        MOV R0, #33H          ; load word address into R0
LOOP:   ADD R0, R1            ; add R1 to R0
        MOV @R0, #00H         ; write a 0 into location pointed to by R0
        INC  R1               ; increment R1
        CJNE R1, #09H, LOOP   ; compare #09H to R1 and jump if not equal
```

Note: Only R1 and R0 can be used for indirect register addressing, as shown earlier in the instruction MOV @R0, #00H.

8051 Flag Bits

The 8051 flag bits are located in the program status word (PSW). The PSW is also called the flag register. There are four conditional flags in the PSW: carry (CY), auxiliary carry (AC), overflow (OV), and parity (P).

Figure 9-8 shows the layout of the PSW word with the four conditional flags: CY, AC, OV, and P. It also has 2 bits which are used to select the register bank to be used. Two remaining bits are available for use by the programmer for any purpose.

The conditional flags are

- **CY (Carry)** The carry flag (CY) is set whenever there is a carry out of bit D7. It is affected by either an 8-bit addition or subtraction. It can also be set directly using single-bit instructions.

- **AC (Auxiliary Carry)** If there is a carry between bits D3 and D4 during an addition or a subtraction, the auxiliary carry (AC) bit is set. It's useful primarily in BCD operations.

- **P (Parity)** The parity flag (P) is basically an odd parity flag. If there is an odd number of 1's in the accumulator, the parity flag is set. If there is an even number of 1's in the accumulator, the parity flag is zero.

- **OV (Overflow)** The overflow flag (OV) is set whenever the result of a signed number operation is too large, causing the high-order bit to overflow into the sign bit. The carry flag is used to detect errors in unsigned arithmetic operations, while the overflow flag is used only to detect errors in signed arithmetic operations.

The instructions that affect the conditional flag bits CY, OV, and AC are shown in Table 9.3. Note that the P flag, the parity bit, is affected by every instruction, so it is not included in the table.

Note that in Table 9.3, X can be a 0 or a 1.

Counter/Timer Programming

The 8051 has two timers—timer0 and timer1—and they can be used as timers or counters. Each timer is 16 bits wide, and therefore is accessed through two separate 8-bit registers: TL0, TH0, and TL1, TH1. TL0 is the low-order 8-bit register for timer 0, while TH0 is the high-order 8-bit register for

Instruction	CY	OV	AC
ADD	X	X	X
ADDC	X	X	X
SUBB	X	X	X
MUL	0	X	
DIV	0	X	
DA	X		
RRC	X		
RLC	X		
SETB C	1		
CLR C	0		
CPL C	X		
ANL C, bit	X		
ANL C, /bit	X		
ORL C, bit	X		
ORL C, /bit	X		
MOV C, bit	X		
CJNE	X		

Table 9.3 ■ Conditional Flags

timer 0. TL1 and TH1 are for timer 1. These registers are SFR registers and can be accessed directly like any other register. For example, "MOV TL1, #33H" puts 33H in the timer1 low-order register.

The TMOD register applies to both timers. It is 8 bits wide and has 4 bits for each timer. The TMOD register sets the modes for both timers. Figure 9-10 shows the layout of the TMOD register.

TMOD Register

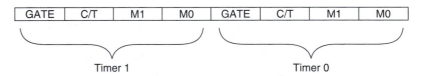

| GATE | C/T | M1 | M0 | GATE | C/T | M1 | M0 |

Timer 1   Timer 0

GATE – When the GATE bit is set, the timer/counter is only enabled while the INTx pin is high and the TRx control pin is set. When the GATE bit is not set, the timer/counter is enabled whenever the TRx control pin is set.

C/T – When the C/T bit is set, it acts as a counter and the input is taken from the Tx input pin. When the C/T bit is not set, it acts as a timer and the input is taken from the internal clock.

M1   M0 – Together these two bits set the mode as shown below.

M1	M0	Mode	Operating Mode
0	0	0	13-bit timer mode
0	1	1	16-bit timer/counter mode
1	0	2	8-bit auto reload mode
1	1	3	Split timer mode

Figure 9-10 ■ The TMOD register layout

Figure 9-10 shows that when the GATE bit for each timer is set, it basically causes the timer/counter to have two enables—the INTx as well as the TRx. If the GATE bit is not set, then the TRx input pin is the only enable for the timer/counter.

The C/T bit determines whether each timer is in the timer or the counter mode. When set, the C/T bit indicates that that timer is in the counter mode, and the input is taken from the Tx input pin. When not set, the timer is in the timer mode and takes the input from the internal system clock.

The M1 M0 bits determine what mode the timer is in. Mode 0 is a 13-bit timer mode, mode 1 is a 16-bit timer mode, and mode 2 is an 8-bit timer. Mode 3, on the other hand, is a split timer mode. Mode 1 and 2 are the most commonly used modes.

The timers count up, as do the counters.

Example 9.9

▼

Problem: Write a program that generates a square wave out on port 0 bit 3 and uses timer 0 to generate the delay used for each pulse.

Solution: The code for this program is shown next.

```
        MOV TMOD, #01H      ; set timer 0 to mode 1 – 16-bit timer mode
OVER:   MOV TL0, #00H       ; load lower byte of timer
        MOV TH0, #FEH       ; load upper byte of timer
        CPL P0.3            ; toggle bit 3 of port 0
        ACALL DELAY         ; call delay subroutine
        SJMP OVER           ; return to do again
DELAY:  SETB TR0            ; start timer 0
DEL2:   JNB TF0, DEL2       ; jump if TF0 is not set
        CLR TR0             ; stop the timer
        CLR TF0             ; clear the timer 0 flag
        RET
```

Note: The timer counts up from FE00H to FFFFH, one count for each clock cycle, and then rolls over to 0000H. When it goes to 0000H, the TF0 flag goes to a 1. This causes the JNB instruction to not transfer and the program falls through to the next instruction.

▲

As one can see, the 8051 microcontroller offers several advantages by having RAM, ROM, and the I/O ports and timers all on one chip. Also, because of this limited configuration, the assembly language for the 8051 often benefits from the fixed configuration and has simpler or more direct instructions. This includes such things as: being able to address RAM with only 8 bits; having so much be single-bit addressable; moving items between RAM, registers, or I/O ports using the same instructions; and being able to directly address the timer/counter registers for setting them or reading them. All of these make programming the 8051 simpler than a general-purpose microprocessor, such as the 8085, once you learn the assembly language of the 8051.

As stated earlier in this chapter, the fixed configuration of the 8051 makes it very desirable for use in embedded systems or fixed applications. However, this fixed configuration does not make it a good fit for an application that requires flexibility or versatility of functions.

Summary

- Microprocessors are general-purpose in nature and hold the microprocessor on a single chip.
- A microcontroller is made up of a microprocessor, RAM, ROM, timers, I/O ports, and serial com ports, all on one chip.
- An embedded system is a system that's implanted in a specific product to do one specific task.
- An embedded system often uses a microcontroller since it is more efficient and cost-effective.
- A microcontroller is a complete system on a single chip. The RAM, ROM, I/O ports, timers, and serial com port are all integrated into the chip with the microprocessor.
- In all microcontrollers, the amount of RAM, ROM, the number of I/O ports, and the timers are all fixed in their quantities and capabilities.
- The 8051 is Intel's 8-bit microcontroller that includes all the common components of a microcontroller in a single chip.
- The 8051 has 4K bytes of ROM, 128 bytes of RAM, two 16-bit timer/event counters, four 8-bit programmable timer/event counters, one serial I/O port, and five interrupt lines.
- Though the 8051 is a self-contained microcontroller, it still has the capability to connect to external ROM and/or RAM.
- The 8051 has 4K bytes of ROM (program memory) and an address spectrum of between 0000H to 0FFFH. This is a separate address space from the data memory.
- RAM is 128 bytes and is divided into register bank 0, register bank 1 (and the stack), register bank 2, register bank 3, bit addressable RAM, and a RAM scratch pad.
- The main registers used in the 8051 are A, B, R0–R7, DPTR (data pointer), SP (stack pointer), and PC (program counter).
- A, B, DPTR, SP, PSW, and PC are SFR registers.
- The PSW register—also called the flag register—is an 8-bit register. It contains four flag bits as well as bits used for selection of the register bank.
- The data transfer instructions move or transfer data in all combinations between registers, RAM, and immediate data.
- The arithmetic operations are Add, Subtract, Increment, Decrement, Multiply, Divide, and Decimal Adjust.
- The logical operations allowed in the 8051 are AND, OR, Exclusive OR, Clear, Complement, Rotate, and Swap.

- The 8051 has the capability to address RAM, registers, I/O ports, and carry using single-bit instructions.
- The branching instructions include unconditional jumps, conditional jumps, calls to subroutines, and returns from subroutines.
- There are four conditional flags in the PSW: carry (CY), auxiliary carry (AC), overflow (OV), and parity (P).
- The 8051 has two timers—timer0 and timer1—which can be used as timers or counters.

Questions

1. What is the basic difference between microprocessors and microcontrollers?
2. Why does a microcontroller fit an embedded system better than a microprocessor?
3. List the components of the 8051.
4. Draw the address spectrum of RAM in the 8051.
5. How is the register bank to use in RAM selected?
6. How does the overflow flag get set?
7. What does the instruction "MOV @Ri, A" do?
8. List the seven arithmetic operations the 8051 carries out.
9. What does the instruction "SWAP A" do?
10. What is the advantage of having bit-addressable instructions?
11. What does the TMOD register do?

Problems

1. Write a program in 8051 assembly language that clears the accumulator, and then adds two to the accumulator nine times.
2. Write a program that adds the two words located in RAM locations 33H and 34H and then stores the answer in RAM location 35H.
3. Write a program that sends out all 1's on port 1 and then delays for a total of FFH cycles. After the delay, send out all 0's on port 1 and delay again.
4. Write a program that monitors bit 0 of port 2, and when it is set, send out 55H on port 3.
5. Write a program to copy ten words in RAM locations starting at 40H, which will then store the ten words in RAM locations starting at 60H.
6. Write a program that reads in three words from RAM locations 7DH, 7EH, and 7FH, finds the average of the three, and stores the result in RAM location 7CH.
7. Write a program that reads in a temperature on port 0 and sees if it is equal to 72D. If it equals 72, put the temperature in register R0. If the temperature is above 72, put the temperature in R1. If the temperature is below 72, put the temperature in R2.

8. Write a program that determines the number of 1's in the data value 77H, and then stores the count of the number of 1's in RAM location 22H.

9. Write a program that checks bit 7 in port 1, and when the bit goes high, set a flag in bit-addressable RAM at location 2CH, and also send out a low on bit 0 in port 2.

10. Write a program that sends out a square wave on bit 1 on port 2, and uses the 8051 timer 0 for the delay of the pulse both on and off. (Set the timer value to 6666H for the delay.)

10

Comparison of the 8085 to Other Microprocessors

Objectives:

Upon completion of this chapter, you should:

- Understand how the 8085 microprocessor compares to the Motorola 6800

- Be able to compare the 8085 with the Motorola 68000

- Know how the 8085 compares to the 8088

- Understand how the 8085 compares to the 80186, 80286, 80386, and 80486

- Comprehend how the 8085 compares to the Pentium

- Understand that knowledge learned about the 8085 microprocessor can be applied to all microprocessors

Key Terms:

- **Direct mode**—Direct addressing in the 6800 when the address is 8 bits wide.

- **Extended mode**—Direct addressing in the 6800, but when the memory address is 16 bits wide

- **Attribute**—.B, .W, and .L that is appended to instructions in the 68000 instruction set to indicate byte, word (16-bit), and long (32-bit)

■ **Segment registers**—8088 registers used by
being added to the pointer registers to form
20-bit memory addresses

Introduction

All microprocessors are similar in many ways. They may differ in the size of
the buses, the number of registers, or the mnemonics used in their assembly language instructions, but basically they are all comparable. In addition, if you learn
how one microprocessor works and how to program it, you can easily learn how
any microprocessor works, and program that, too. As for programming, all assembly languages are at the same level, and basically perform the same operations.

The following sections highlight the differences (as well as the similarities) between the 8085 and several other microprocessors.

10-1 The 8085 Compared to the 6800

The Motorola 6800 is an 8-bit microprocessor that came out at about the
same time the 8085 did. It is comparable to the 8085 in many ways, but there
are also disparities in the hardware and software.

Hardware Comparison to the 6800

The 6800 is a 40-pin dip package, the same as the 8085, and is considered
to be an 8-bit microprocessor since the data bus is 8 bits wide. The address bus
is 16 bits wide, like the 8085, except the 6800 does not have a multiplexed bus.
Figure 10-1 shows the pinout of the 6800.

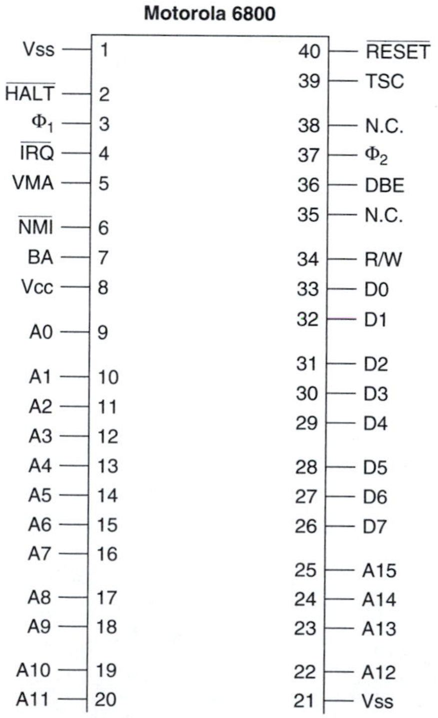

Figure 10-1 ■ The Motorola 6800 pinout

As shown in Figure 10-1, the 16 bits of the address bus and the 8 bits of the data bus are not multiplexed. This makes hardware connections easier than the 8085 and its multiplexed bus. Since the address bus is 16 bits wide, it allows the 6800 to access 2^{16} bytes of memory, or 64K bytes—the same as the 8085.

The 6800 runs at 2MHz, which is essentially the same clock speed as the 8085, and uses a single five-volt supply voltage.

Figure 10-2 shows a functional block diagram of the 6800.

As displayed in Figure 10-2, the 6800 has two interrupts, a regular interrupt and a nonmaskable interrupt. This is another area where the 8085 differs in that it has five interrupts, four maskable interrupts, and one nonmaskable.

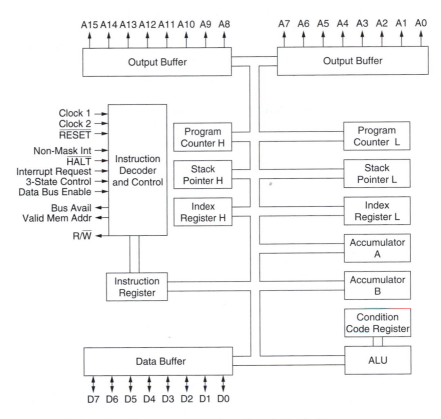

Figure 10-2 ■ The Motorola 6800 functional block diagram

The 6800 also has two 8-bit accumulators and one index register. The other three registers are the stack pointer, the program counter, and the condition code register. This differs from the 8085 which has only one accumulator, but has six general-purpose registers: B, C, D, E, H, and L. The 8085 also has special-purpose registers similar to the 6800 (stack pointer, program counter, and flag register). The 6800's condition code register contains the flags and is what the 8085 calls the flag register.

Software Comparison to the 6800

The instruction set of the 6800 contains 72 basic instructions. This set does not include any I/O instructions like IN and OUT in the 8085. Instead, the 6800 only allows and uses memory-mapped I/O. This denotes where the I/O addresses are in the memory spectrum, meaning the I/O addresses are not their own address range. This is significantly different from the 8085 which allows for both peripheral-mapped I/O and memory-mapped I/O. As for the number of instructions, the 8085 has 74—basically the same number as the 6800.

The 6800 has one-, two-, and 3-byte instructions—the same as the 8085—which are grouped into the following categories:

- Accumulator and memory instructions
- Index register and stack pointer instructions
- Jump and branch instructions
- Condition code register instructions

In the 8085, the same instruction may have different op codes based upon the register used, not the addressing modes. For different addressing modes in the 8085, different instruction mnemonics are employed.

Some mnemonics in the 6800 have as many as four possible op codes, depending upon the addressing modes. The addressing modes in the 6800 are immediate, direct, indexed, and extended. In the 6800, the **direct mode** and the **extended mode** are both what the 8085 calls direct addressing. In the 6800, direct addressing only applies when the memory address is 8 bits wide, and therefore is in the beginning section of memory. Extended addressing is direct addressing, but in this case the memory address is 16 bits wide.

For example, the instruction "LDAA" loads memory into accumulator A.

Table 10.1 shows that the actual op code for the instruction LDAA varies based upon which type of addressing is being used.

Mnemonic	Operand	Op Code	Addressing Mode	Comment
LDAA	#$F3	86	Immediate	Load A with F3H
LDAA	$F3	96	Direct	Load A with data from memory location F3H
LDAA	$3,X	E6	Indexed	Load A with data from memory location (3+contents of index register)
LDAA	$0112	F6	Extended	Load A with data from memory location 0112H

Table 10.1 ■ Instruction LDAA

Note that a $ sign is used to indicate that the number is in hexadecimal, and the # sign is employed to indicate immediate addressing. Also, "X" means indexed addressing using the index register. As mentioned previously, the direct mode in the 6800 only refers to direct addressing with 8-bit wide addresses. Extended addressing is direct addressing but with 16-bit wide addresses.

The form of the instructions in the 6800 is similar to the 8500, as shown next:

```
Label    mnemonic    operand    comment
Start    STAB        $0122      store accumulator B in memory at location 0122H
```

Note that there is no ":" after the label Start or ";" before the comment as happens in 8085 code.

10-2 The 8085 Compared to the 68000

Motorola's 68000 microprocessor was developed after the 6800. The 68000 is considered a 16-bit microprocessor since it has a 16-bit wide data bus. But since internally it contains 32-bit wide registers, it is sometimes referred to as a 16-bit external/32-bit internal microprocessor. It also has a 24-bit wide address bus.

Hardware Comparison to the 68000

The 68000 is available in a 64-pin DIP package, and is considered a 16-bit microprocessor since it has a 16-bit data bus. But, as mentioned, internally it has a 32-bit wide register, and a 24-bit address bus that is capable of addressing 2^{24} bytes (16MB). It does not have a multiplexed bus.

So, as compared to the 8085, the data bus is twice as wide, the address bus and resulting address spectrum are larger (8085—16-bit address bus; 68000—24-bit address bus). And, of course, the 8085 has a multiplexed bus.

Figure 10-3 shows the input and output signals for the 68000.

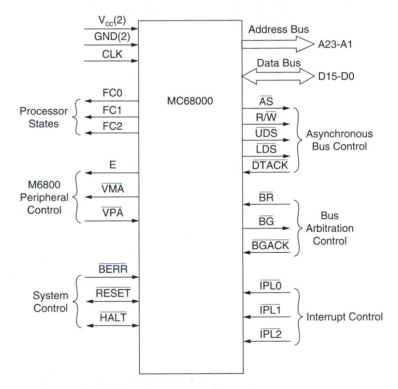

Figure 10-3 ■ Motorola 68000 input/output signals

As shown in Figure 10-3, the data bus is 16 bits wide and bidirectional, while the address bus is 24 bits wide but only A1-A23 is accessible. A0 is not available because the 68000 always accesses instructions and data from even-numbered addresses.

The asynchronous bus control signals are for handling asynchronous data transfers. The signals are address strobe ($\overline{AS}$), read/write (R/$\overline{W}$), upper and lower data strobe ($\overline{UDS}$, $\overline{LDS}$), and data transfer acknowledge ($\overline{DTACK}$).

The Bus Arbitration Control signals are used in determining who is going to be the bus master. The signals are Bus Request ($\overline{BR}$), Bus Grant ($\overline{BG}$), and the Bus Grant Acknowledge ($\overline{BGACK}$).

The Interrupt Control signals are used to determine the priority level of the device requesting an interrupt. The signals include Interrupt Priority Levels 0, 1, and 2 ($\overline{IPL0}$, $\overline{IPL1}$, $\overline{IPL2}$).

The System Control signals are used to reset or halt the microprocessor. The signals are Bus Error ($\overline{BERR}$), Reset ($\overline{RESET}$), and Halt ($\overline{HALT}$).

The MC6800 Peripheral Control signals are used to interface synchronous 6800 peripherals with the asynchronous 68000. The signals include Enable (E), Valid Peripheral Address ($\overline{VPA}$), and Memory Address ($\overline{VMA}$).

The Process Status signals indicate the state and cycle type being executed. The signals are Function Code 0, 1, and 2 (FC0, FC1, FC2).

These signals, although they often have different acronyms and slightly different functions, are pretty much the same for all microprocessors. Besides power, ground, and a clock, there is the data bus, the address bus, and then assorted control signals. The control signals usually include some sort of interrupts, halt and reset signals, an address latch or signal, read/write signals, memory versus I/O (when I/O instructions are used), and function or state signals.

The 8085 has similar control signals to the 68000, except it has an IO/$\overline{M}$ signal which the 68000 does not because it only allows memory-mapped I/O. Also, the 8085 has Serial Input and Output (SID, SOD), which the 68000 does not have. However, the 68000 has control signals to synchronize with 6800 peripherals, while the 8085 does not have any comparable signals.

Also, the 68000 clock speed started at 4MHz, but now has versions at 6, 8,10, and 12MHz. The 8085's clock speed is 3MHz.

Software Comparison to the 68000

Probably the biggest difference between the 68000 instructions and the 8085 instructions are the format with regards to the source and destination. In the 68000 instruction set, the format is instruction, source, destination, while the 8085's format is instruction, destination, source.

When considering the instruction set of any microprocessor, you should consider the Programmers Model. Figure 10-4 shows the Programmers Model for the 68000.

As shown in Figure 10-4, the 68000 has eight 32-bit data registers, eight 32-bit address registers, a 32-bit program counter, and a 16-bit status register (flags).

The eight data registers are D0–D7, which can be used on a byte basis, a word basis (16 bit), and a long word basis (32 bit). Operands in the 68000 instruction set can use an **attribute** of .B, .W, or.L to denote byte, word , or long word. For example, the instruction:

```
MOVE.B #$7F, D2
```

moves the byte 7FH into the low-order byte of the data register D2. The remaining bits in register D2 are undisturbed. The default case of the MOVE instructions is .W or 16 bits. So one does not need to use MOVE.W, only MOVE.B, and MOVE.L.

Note that like the 6800, the # sign indicates immediate addressing, while the $ sign indicates hexadecimal.

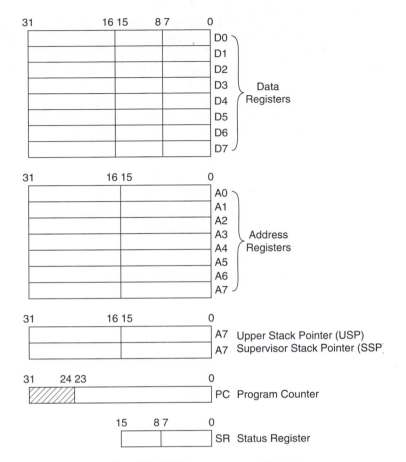

Figure 10-4 ■ The 68000 Programmers Model

Many 68000 instructions use the eight address registers and the eight data registers in the same fashion, while some instructions use the address registers only.

Also, byte operations are not permitted on address registers. When the .W is used in an address register, the value is extended out to 32 bits and is sign-extended. For example, the instruction:

MOVEA   #$2000, A5

moves the value $00002000 into address register A5. When the address registers are used to refer to physical memory locations, only 24 bits are used since that is the size of the address bus.

The program counter (PC) is 32 bits wide but only 24 bits are used. It's used as in the 8085 and most microprocessors to store the address of the next instruction to be executed.

The stack pointer (SP) is address register A7 in the 68000. Which stack pointer is being used (user or supervisor) is determined by a bit in the status register. Normally, the user mode is used, and the User Stack pointer is the stack pointer. (When the system is put in the supervisor mode, the supervisor stack pointer is used.) In instructions, one can use either SP or A7 to reference the stack pointer. The stack in the 68000 grows downwards, instead of up, as in the 8085 stack. In this situation, the stack is used (as in the 8085) for saving the return address when going to a subroutine, and can also be addressed directly.

The status register (SR) is grouped into the User byte—in the lower 8 bits, and the System byte—in the upper 8 bits. The user byte has the flags, while the system byte has the interrupt mask and the supervisor bit.

In the 68000, 13 addressing modes are used, while the 8085 has four addressing modes. The 68000 addressing modes are shown next:

- Data register direct
- Address register direct
- Absolute short
- Absolute long
- Register Indirect
- Postincrement register indirect
- Predecrement register indirect
- Register indirect with offset
- Register indirect with index and offset
- PC-relative with offset
- PC-relative with Index and offset
- Immediate
- Implied registers

This many addressing modes make for a more involved instruction set, but the same basic types of instructions exist and are performed.

The 68000 instructions are grouped into the following types:

- Data movement
- Integer arithmetic
- Boolean
- Shift and rotate
- Bit manipulation
- Binary-coded decimal
- Program flow
- System control

These groups are basically the same for all microprocessors. Some microprocessors—like the 8085—use a fewer number of groups (five for the 8085), but these groups are larger. The 68000 just breaks up a few of the 8085 instruction sets into two or three groups.

Also, for ease of learning, the 68000 instruction set can be broken up into three groups:

- Introductory instructions
- Intermediate instructions
- Advanced instructions

The introductory group contains the basic instructions that are common to most 8-bit microprocessors. The intermediate group contains slightly more complex instructions that a lot of 8-bit microprocessors don't have. The final group—advanced—contains the most advanced instructions the

68000 has, but one doesn't need to use any of the advanced instructions when programming the 68000.

10-3 The 8085 Compared to the 8088

The 8088 was developed by Intel after the 8086, which was designed to be forward compatible with the older 8085 (meaning programs written for the 8085 will run on the 8086). The 8088 is considered by many to be a 16-bit microprocessor, even though it has only an 8-bit data bus. (The 8086, by contrast, has a 16-bit data bus.) The 8088 is considered to be a 16-bit microprocessor because it has 16-bit internal registers. The 8088 also has a 20-bit address bus, which is multiplexed with the data bus and some status signals.

The Hardware Comparison with the 8088

The 8088 uses a 5V supply, comes in a 40-pin DIP package, and runs at a clock speed of 5MHz. The data bus is 8 bits wide, and the address bus is 20 bits wide, which allows the 8088 to address 2^{20} bytes or 1MB. The low-order 8 bits of the address bus are multiplexed with the data bus, while the high-order 4 bits of the address bus are multiplexed with status information.

Figure 10-5 shows the pinout of the 8088.

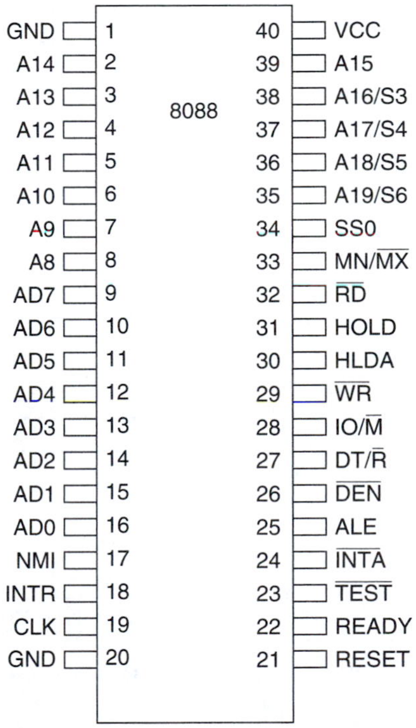

```
        GND ▭ 1          40 ▭ VCC
        A14 ▭ 2          39 ▭ A15
        A13 ▭ 3    8088  38 ▭ A16/S3
        A12 ▭ 4          37 ▭ A17/S4
        A11 ▭ 5          36 ▭ A18/S5
        A10 ▭ 6          35 ▭ A19/S6
         A9 ▭ 7          34 ▭ SS0
         A8 ▭ 8          33 ▭ MN/M̄X̄
        AD7 ▭ 9          32 ▭ R̄D̄
        AD6 ▭ 10         31 ▭ HOLD
        AD5 ▭ 11         30 ▭ HLDA
        AD4 ▭ 12         29 ▭ W̄R̄
        AD3 ▭ 13         28 ▭ IO/M̄
        AD2 ▭ 14         27 ▭ DT/R̄
        AD1 ▭ 15         26 ▭ D̄ĒN̄
        AD0 ▭ 16         25 ▭ ALE
        NMI ▭ 17         24 ▭ ĪN̄T̄Ā
       INTR ▭ 18         23 ▭ T̄ĒS̄T̄
        CLK ▭ 19         22 ▭ READY
        GND ▭ 20         21 ▭ RESET
```

Figure 10-5 ■ An 8088 pinout diagram

As shown in Figure 10-5, the data bus is 8 bits wide and is multiplexed with the lower 8 bits of the address bus (AD0–AD7), while the address bus is 20 bits wide (AD0–AD7, A8–A19). The upper 4 bits of the address bus A16–A19 are multiplexed with the state signals S3–S6.

Many of the control signals are the same as the 8085, but some differences exist. For instance, the 8085 has RST 5.5–7.5 for interrupts, but the 8088 does not have a counterpart for them. The 8088 has INTR and NMI interrupt inputs, both of which the 8085 has (the 8085's TRAP is the same as the NMI input). The 8088 also has control signals $\overline{\text{DEN}}$ and DT/$\overline{\text{R}}$, which are used to control external data-bus buffers when needed. The 8085 has no comparable signals.

Basically, there are a lot of similarities hardware-wise between the 8088 and the 8085. This is mainly because they are both Intel processors, and the 8088 was designed to be upward compatible with the 8085.

Software Comparison to the 8088

The software for the 8088 is obviously going to be very similar to the software for the 8085. Since it is an Intel microprocessor, it has the same form as the 8085: instruction, destination, source. But the 8088 (and the 8086) have a larger instruction set—135 basic instructions compared to the 8085's set of 74 instructions.

Figure 10-6 shows the programming model for the 8088.

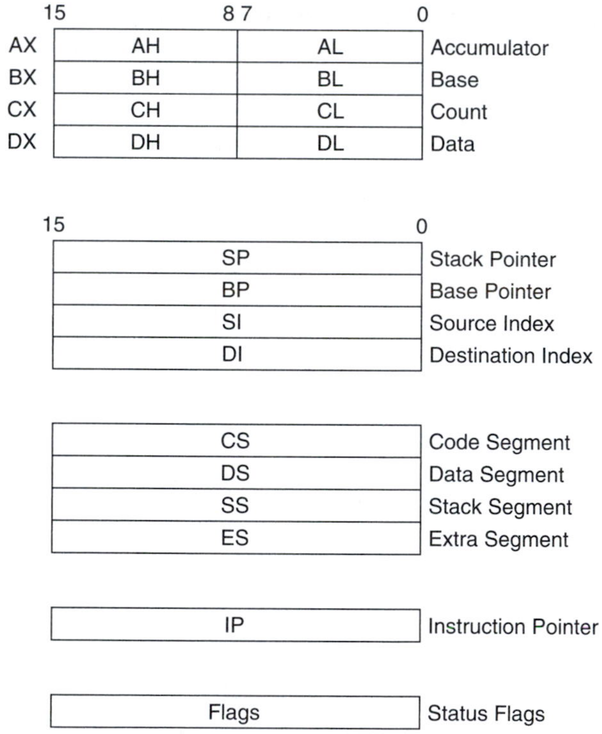

	15	8 7	0	
AX	AH	AL		Accumulator
BX	BH	BL		Base
CX	CH	CL		Count
DX	DH	DL		Data

15	0	
SP		Stack Pointer
BP		Base Pointer
SI		Source Index
DI		Destination Index

CS	Code Segment
DS	Data Segment
SS	Stack Segment
ES	Extra Segment

IP	Instruction Pointer

Flags	Status Flags

Figure 10-6 ■ The 8088 programming model

As seen in Figure 10-6, the 8088 has four 16-bit general-purpose registers: AX, BX, CX, and DX. Any of these four registers can be used as an accumulator, but AX is most often employed as the primary accumulator. BX is often used as a base register, CX as a loop counter, and DX as an I/O address pointer or data register. These four general-purpose registers can also be used as sets of 8-bit register pairs (see Figure 10-6).

Other registers shown in Figure 10-6 are the stack pointer (SP), the base address pointer (BP), the source index pointer (SI), and the destination index pointer (DI). These registers are all used as memory pointers. The 8085, on the other hand, does not have any of these memory pointer registers.

The **segment registers:** the code segment (CS), data segment (DS), stack segment (SS), and the extra segment (ES) are all used to make up the 20-bit effective address by being added to the 16-bit pointer registers to make up 20-bit memory addresses.

The remaining registers are the instruction pointer (IP) and the status register. The status register contains the flags. The 8088 uses the following nine addressing modes:

- Immediate addressing
- Register addressing
- Direct addressing
- Register indirect addressing
- Based addressing
- Indexed addressing
- Based indexed addressing
- String addressing
- I/O addressing

Recall that the 8085 uses four addressing modes. The 8088 instructions can be sorted into the following groups:

- Data transfer
- Arithmetic
- Logic
- Program transfer
- String manipulation
- Machine or processor control

All of these categories of instructions exist in the 8085 except for the string manipulation group of instructions. This group is new to the 8088.

The 8088 instructions can operate on individual bits, bytes, 16-bit words, and 32-bit double words, signed numbers, ASCII characters, and BCS numbers.

10-4 The 8085 Compared to the 80186 and 80286

Intel developed the 80186 as a 16-bit microprocessor that is an extended version of the 8086. One major change was that the 80186 was put in a 68-pin PLCC package instead of the 40-pin DIP package. The 80186 also uses some new processor techniques such as prefetched pipeline structure, parallel processing, and memory management.

The 80186 is basically an improved version of the 8086 that runs at either 8MHz or 10MHz. It still has multiplexed data and address buses, and the additional lines of the larger pin package were used to include devices such as a clock generator, an interrupt controller, timers, a DMA controller, and a chip select unit. The 80186 can address 1MB of memory over its 20-bit address bus.

The 80186 uses the exact same register set as the 8086/8088. Regarding the instruction set, the 80186 is upward compatible with the 8086/8088, but has added 10 additional instructions over the 8088 instruction set.

The 80286 is also a 16-bit microprocessor, but it's an improved version of the 8086. Like the 80186, it too comes in a 68-pin PLCC package. However, the 80286 has used a different architectural philosophy. The multiplexed buses were eliminated, and the address bus is 24 bits wide and can address 16MB of memory. The 80286 also can additionally address 1GB of memory through a virtual memory management system.

The 80286 instruction set is the same as the 80186 instruction set, except it has an additional 16 instructions. However, the 80286 is also software upward compatible with the 8086/8088.

10-5 The 8085 Compared to the 80386 and 80486

The 80386 microprocessor by Intel is a 32-bit microprocessor with 32-bit internal registers, a 32-bit wide data bus, and a 32-bit wide address bus. It is packaged in a 132-pin grid array package and can operate at speeds ranging from 20MHz to 33MHz. It can also address 4GB of physical memory, as well as 64TB of virtual memory.

Figure 10-7 shows the programming model for the 80386.

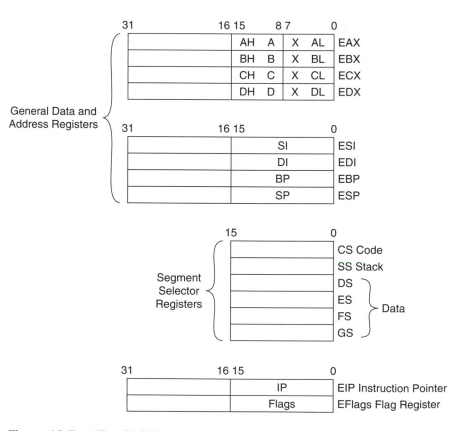

Figure 10-7 ■ The 80386 programming model

Figure 10-7 shows that the registers are basically 32-bit versions of the 8088 registers. The 80386 has 32-bit registers—EAX, EBX, ECX, and EDX—but can also address 16-bit registers as AX, BX, CX, and DX. The segment registers are basically the same as in the 8088—as are the instruction pointer and the flag register.

Basically, the programming model for the 80386 is an expanded version of the 8088 programming model.

The instruction set of the 80386 has 11 addressing modes and is organized into nine groups of instructions. The instruction set contains all of the basic 8085 type of operations, but it also has instructions that do string manipulation, bit manipulation, high-level language support, and operating system support.

The 80486 is an upgraded version of the 80386, and is packaged in a 168-pin grid array, operating at frequencies from 25MHz to 66MHz. It is considered to be a high-speed, high-performance 32-bit microprocessor. Some of the major differences between the 80486 and the 80386 are that the following items are on the 80486 chip:

- A built-in math coprocessor
- 8Kbytes of data and code cache memory
- A highly pipelined execution unit

The 80486 was used in many PCs and networking applications. Although it is a fairly involved microprocessor, at the heart of this and other microprocessors is still the 8-bit microprocessor you learned about earlier.

10-6 The 8085 Compared to the Pentium

The Intel Pentium microprocessor is upward compatible from the 80486 and goes all the way back to the 8088 in terms of software compatibility. It is currently the microprocessor used in many PCs. The Pentium has a 64-bit data bus, a 32-bit address bus, and can run at clock speeds ranging from 60MHz to 233MHz. It is packaged in a 273-pin grid array. The 64-bit data bus allows the Pentium to increase its processing speed.

The Pentium added several advanced features, such as the following:

- Superscalar architecture
- On-chip cache memory for data and code
- Branch prediction
- A high-performance floating-point unit
- Performance monitoring

All of these advanced features are aimed at increasing the performance of the Pentium.

There are many variations of the Pentium processor (Pentium I, Pentium II, Pentium III, Pentium Pro), all with slightly different capabilities, speeds, and capacities. But they are all part of the Pentium family of microprocessors. Though the Pentium microprocessor is a bit more advanced than the 8085 microprocessor, it is still basically a microprocessor. And, given the specification sheets, the programming model, and the instruction set, you could design and program a basic microprocessor system using a Pentium.

Summary

- The 6800 is an 8-bit microprocessor in a 40-pin DIP package. Its data bus is 8 bits wide, while the address bus is 16 bits.

- The instruction set of the 6800 contains 72 basic instructions, but does not include any I/O instructions. Instead, the 6800 only allows and uses memory-mapped I/O.

- One major difference between the 6800 instruction set and the 8085's is that the 6800 uses different op codes for the same instruction based upon the addressing mode used.

- The 68000 is considered a 16-bit microprocessor since it has a 16-bit wide data bus. Since it has 32-bit wide registers internally, it is sometimes referred to as a 16-bit external/32-bit internal microprocessor.

- The 68000 has a 24-bit address bus that is capable of addressing 2^{24} bytes (16MB). It does not have a multiplexed bus.

- In the 68000 instruction set, the format is instruction, source, destination. In contrast, the 8085's format is instruction, destination, source.

- The 8088 is considered to be a 16-bit microprocessor because it has 16-bit internal registers. The 8088 also has an 8-bit data bus and a 20-bit address bus, both of which are multiplexed with the data bus and some status signals.

- The 8088 can address 2^{20} bytes or 1MB of memory.

- There are a lot of similarities between the 8088 and the 8085. This is because they are both Intel processors, and the 8088 was designed to be upward compatible with the 8085.

- The 8088 (and the 8086) have a larger instruction set: 135 basic instructions as compared to the 8085's instruction set of 74 instructions.

- The 8088 instruction set is very similar to the 8085 instruction set.

- The 80186 is an improved version of the 8086 that runs at either 8MHz or 10MHz.

- The 80186 still has multiplexed data and address buses, like the 8085.

- The 80186 can address 1MB of memory over its 20-bit address bus.

- The 80286 is a 16-bit microprocessor, and is also an improved version of the 8086.

- The 80286 comes in a 68-pin PLCC package.

- The 80286 does not have multiplexed buses, and has a 16-bit data bus and a 24-bit address bus.

- The 80286 can address 16MB of memory, and can additionally address 1GB of memory through a virtual memory management system.

- The 80386 microprocessor is a 32-bit microprocessor with 32-bit internal registers, a 32-bit wide data bus, and a 32-bit wide address bus.

- The 80486 is an upgraded version of the 80386 and is packaged in a 168-pin grid array, operating at frequencies ranging from 25MHz to 66MHz.
- The Pentium has a 64-bit data bus and a 32-bit address bus, and can run at clock speeds from 60MHz up to 233MHz. It is packaged in a 273-pin grid array.

Questions

1. How does the 6800 differ from the 8085 in terms of interrupts?
2. How does the 6800 refer to the accumulator when using it in an instruction?
3. What is unique about the 6800 with regards to I/O instructions?
4. How do the Motorola processors indicate that the data is immediate?
5. Why is the Motorola 68000 sometimes called a 16/32 microprocessor?
6. How does the instruction format of the 68000 compare to the instruction format of the 8085?
7. What are the addressing modes used in the 68000?
8. Why is the 8088 considered a 16-bit microprocessor even though it has an 8-bit data bus?
9. What signals are multiplexed in the 8088?
10. In the 8088, which register(s) can be used as the accumulator?
11. Why does the 80186 still have a multiplexed bus even though it has more pins than the 8088?
12. In the 80286, what has been eliminated from the 80186?
13. What are the major architectural changes added to the 80486 microprocessor over the 80386?
14. What are the basic sizes and widths of the Pentium processor?
15. What major architectural changes have been made to the Pentium processor?

Appendices

Appendix I—The 8085 Instruction Set

ACI data(8b)

> **Description:** Add Immediate 8-bit data to the accumulator with carry.
>
> **Bytes/M-Cycles/T-States:** 2/2/7
>
> **Hex Code:** CE
>
> **Flags:** All flags are affected based upon the result of the addition.

ADC R

> **Description:** Add register to accumulator with carry.
>
> **Bytes/M-Cycles/T-States:** 1/1/4

	Register
Hex Codes: 8F	A
88	B
89	C
8A	D
8B	E
8C	H
8D	L

> **Flags:** All flags are affected based upon the result of the addition.

ADC M

> **Description:** Add contents of memory location pointed to by HL register pair to the accumulator with carry.
>
> **Bytes/M-Cycles/T-States:** 1/2/7
>
> **Hex Codes:** 8E
>
> **Flags:** All flags are affected based upon the result of the addition.

ADD R

> **Description:** Add register to accumulator.
>
> **Bytes/M-Cycles/T-States:** 1/1/4

	Register
Hex Codes: 87	A
80	B
81	C
82	D
83	E
84	H
85	L

> **Flags:** All flags are affected based upon the result of the addition.

ADD M

> **Description:** Add contents of memory location pointed to by HL to the accumulator.
>
> **Bytes/M-Cycles/T-States:** 1/2/7
>
> **Hex Codes:** 86
>
> **Flags:** All flags are affected based upon the result of the addition.

ADI data(8b)

> **Description:** Add the immediate 8-bit data to the accumulator.
>
> **Bytes/M-Cycles/T-States:** 2/2/7
>
> **Hex Codes:** C6
>
> **Flags:** All flags are affected based upon the result of the addition.

ANA R

> **Description:** The contents of the accumulator and the register are logically ANDed and the result is put in the accumulator.
>
> **Bytes/M-Cycles/T-States:** 1/1/4

	Register
Hex Codes: A7	A
A0	B
A1	C
A2	D
A3	E
A4	H
A5	L

> **Flags:** S, Z, and P are modified based upon the result of the operation. CY is reset, and AC is set.

ANA M

> **Description:** The contents of the accumulator and the contents of the memory location pointed to by HL are logically ANDed, and the result is put in the accumulator.
>
> **Bytes/M-Cycles/T-States:** 1/2/7
>
> **Hex Codes:** A6
>
> **Flags:** S, Z, and P are modified based upon the result of the operation. CY is reset, and AC is set.

ANI data(8b)

> **Description:** The contents of the accumulator and the 8-bit data are ANDed and the result is put in the accumulator.
>
> **Bytes/M-Cycles/T-States:** 2/2/7
>
> **Hex Codes:** E6
>
> **Flags:** S, Z, and P are modified based upon the result of the operation. CY is reset, and AC is set.

CALL address(16b)

> **Description:** The program sequence is transferred to the address specified by the 16-bit address. Before the program is transferred, the address of the instruction following the CALL instruction is pushed onto the stack.
>
> **Bytes/M-Cycles/T-States:** 3/5/18
>
> **Hex Codes:** CD
>
> **Flags:** No flags are affected.

CC address(16b)

> **Description:** The program sequence is transferred to the address specified by the 16-bit address if the CY flag is set. If CY = 0, no transfer takes place. If the transfer takes place, the address of the instruction following the CC instruction is pushed onto the stack.
>
> **Bytes/M-Cycles/T-States:**   3/2/9   if transfer is not taken
>   3/5/18   if the transfer is taken
>
> **Hex Codes:** DC
>
> **Flags:** No flags are affected.

CNC address(16b)

> **Description:** The program sequence is transferred to the address specified by the 16-bit address if the CY flag is not set. If CY = 1, no transfer takes place. If the transfer takes place, the address of the instruction following the CNC instruction is pushed onto the stack.
>
> **Bytes/M-Cycles/T-States:**   3/2/9   if transfer is not taken
>   3/5/18   if the transfer is taken
>
> **Hex Codes:** D4
>
> **Flags:** No flags are affected.

CP address(16b)

> **Description:** The program sequence is transferred to the address specified by the 16-bit address if positive, or if the S flag = 0. If S = 1, no transfer takes place. If the transfer takes place, the address of the instruction following the CP instruction is pushed onto the stack.
>
> **Bytes/M-Cycles/T-States:**   3/2/9   if transfer is not taken
>   3/5/18   if the transfer is taken
>
> **Hex Codes:** F4
>
> **Flags:** No flags are affected.

CM address(16b)

> **Description:** The program sequence is transferred to the address specified by the 16-bit address if minus, or if the S flag = 1. If S = 0, no transfer takes place. If the transfer takes place, the address of the instruction following the CM instruction is pushed onto the stack.
>
> **Bytes/M-Cycles/T-States:**   3/2/9   if transfer is not taken
>   3/5/18   if the transfer is taken
>
> **Hex Codes:** FC
>
> **Flags:** No flags are affected.

CPE address(16b)

> **Description:** The program sequence is transferred to the address speci-
> fied by the 16-bit address if parity is even, or the P flag = 1. If P = 0,
> no transfer takes place. If the transfer takes place, the address of the
> instruction following the CPE instruction is pushed onto the stack.
>
> **Bytes/M-Cycles/T-States:** 3/2/9 if transfer is not taken
>
> 3/5/18 if the transfer is taken
>
> **Hex Codes:** EC
>
> **Flags:** No flags affected.

CPO address(16b)

> **Description:** The program sequence is transferred to the address speci-
> fied by the 16-bit address if parity is odd, or if the P flag = 0. If P = 1,
> no transfer takes place. If the transfer takes place, the address of the
> instruction following the CPO instruction is pushed onto the stack.
>
> **Bytes/M-Cycles/T-States:** 3/2/9 if transfer is not taken
>
> 3/5/18 if the transfer is taken
>
> **Hex Codes:** E4
>
> **Flags:** No flags are affected.

CZ address(16b)

> **Description:** The program sequence is transferred to the address
> specified by the 16-bit address if zero, or if the Z flag = 1. If Z = 0, no
> transfer takes place. If the transfer takes place, the address of the
> instruction following the CZ instruction is pushed onto the stack.
>
> **Bytes/M-Cycles/T-States:** 3/2/9 if transfer is not taken
>
> 3/5/18 if the transfer is taken
>
> **Hex Codes:** CC
>
> **Flags:** No flags are affected.

CNZ address(16b)

> **Description:** The program sequence is transferred to the address speci-
> fied by the 16-bit address if not zero, or if the Z flag = 0. If Z = 1, no
> transfer takes place. If the transfer takes place, the address of the
> instruction following the CNZ instruction is pushed onto the stack.
>
> **Bytes/M-Cycles/T-States:** 3/2/9 if transfer is not taken
>
> 3/5/18 if the transfer is taken
>
> **Hex Codes:** C4
>
> **Flags:** No flags are affected.

CMA

> **Description:** The contents of the accumulator are complemented.
>
> **Bytes/M-Cycles/T-States:** 1/1/4
>
> **Hex Codes:** 2F
>
> **Flags:** No flags are affected.

CMC

Description: The carry flag is complemented.

Bytes/M-Cycles/T-States: 1/1/4

Hex Codes: 3F

Flags: The CY flag is complemented. No other flags are affected.

CMP R

Description: The contents of the register are compared to the contents of the accumulator. Both contents are unaffected, and the following flags are used to show the results of the compare:

If A < R   CY = 1 and Z = 0

If A = R   CY = 0 and Z = 1

If A > R   CY = 0 and Z = 0

Bytes/M-Cycles/T-States: 1/1/4

Hex Codes:	Registers
BF	A
B8	B
B9	C
BA	D
BB	E
BC	H
BD	L

Flags: S, P, and AC are also affected based upon the results of the operation, besides Z and CY.

CMP M

Description: The contents of the memory location pointed to by HL are compared to the contents of the accumulator. Both contents are unaffected, and the following flags are used to show the results of the compare:

If A < R   CY = 1 and Z = 0

If A = R   CY = 0 and Z = 1

If A > R   CY = 0 and Z = 0

Bytes/M-Cycles/T-States: 1/2/7

Hex Codes: BE

Flags: S, P, and AC are also affected based upon the results of the operation, besides Z and CY.

CPI data(8b)

Description: The 8-bit data is compared with the contents of the accumulator. The contents of the accumulator are unaffected. The following flags are used to show the results of the compare:

If A < R   CY = 1 and Z = 0

If A = R   CY = 0 and Z = 1

If A > R   CY = 0 and Z = 0

Bytes/M-Cycles/T-States: 2/2/7

Hex Codes: FE

Flags: S, P, and AC are also affected based upon the results of the operation, besides Z and CY.

DAA

Description: The contents of the accumulator are converted from a binary value to two 4-bit Binary Coded Decimal (BCD) digits.

Bytes/M-Cycles/T-States: 1/1/4

Hex Codes: 27

Flags: S, Z, AC, P, and CY flags are affected based upon the results of the operation.

DAD Rp

Description: The contents of the register pair (Rp) are added to the contents of the register pair HL. The source register pair is unchanged, and the results are stored in HL.

Bytes/M-Cycles/T-States: 1/3/10

		Register Pair
Hex Codes:	09	BC
	19	DE
	29	HL
	39	SP

Flags: If the result is larger than 16 bits, the CY flag is set, otherwise no flags are affected.

DCR R

Description: The contents of the register are decremented by one. The result is stored in the register.

Bytes/M-Cycles/T-States: 1/1/4

		Register
Hex Codes:	3D	A
	05	B
	0D	C
	15	D
	1D	E
	25	H
	2D	L

Flags: S, AC, Z, and P are affected by the results of the operation. The CY flag is not affected.

DCR M

Description: The contents of the memory location pointed to by HL is decremented by one, and the results are stored in the memory location.

Bytes/M-Cycles/T-States: 1/3/10

Hex Code: 35

Flags: S, AC, Z, and P are affected by the results of the operation. The CY flag is not affected.

DCX Rp

> **Description:** The contents of the register pair are decremented by 1. The result is stored in the register pair. The register pair is treated as a 16-bit number.
>
> **Bytes/M-Cycles/T-States:** 1/1/6
>
> Register Pair
>
> | **Hex Codes:** | | |
> |---|---|---|
> | 0B | BC |
> | 1B | DE |
> | 2B | HL |
> | 3B | SP |
>
> **Flags:** No flags are affected.

DI

> **Description:** The Interrupt Enable flip-flop is reset, and all of the interrupts except the TRAP interrupt are disabled.
>
> **Bytes/M-Cycles/T-States:** 1/1/4
>
> **Hex Codes:** F3
>
> **Flags:** No flags are affected.

EI

> **Description:** The interrupt Enable flip-flop is set and all interrupts are enabled.
>
> **Bytes/M-Cycles/T-States:** 1/1/4
>
> **Hex Codes:** FB
>
> **Flags:** No flags are affected.

HLT

> **Description:** The MPU finishes executing the current instruction and halts any further execution. The MPU enters the Halt Acknowledge machine cycle, and Wait states are inserted in every clock period. It requires an interrupt or a reset to get the MPU out of the Halt state.
>
> **Bytes/M-Cycles/T-States:** One / two or more / five or more
>
> **Hex Codes:** 76
>
> **Flags:** No flags are affected.

IN port address(8b)

> **Description:** The contents of the input port designated are read and loaded into the accumulator.
>
> **Bytes/M-Cycles/T-States:** 2/3/10
>
> **Hex Codes:** DB
>
> **Flags:** No flags are affected.

INR R

> **Description:** The contents of the register are incremented by one and stored in the register.
>
> **Bytes/M-Cycles/T-States:** 1/1/4
>
> Register
>
> **Hex Codes:** 3C   A
> 04   B
> 0C   C
> 14   D
> 1C   E
> 24   H
> 2C   L
>
> **Flags:** S, Z, P, AC are affected by the results of the operation. CY is not modified.

INR M

> **Description:** The contents of the memory location pointed to by HL are incremented by one and the result is put in the memory location.
>
> **Bytes/M-Cycles/T-States:** 1/3/10
>
> **Hex Codes:** 34
>
> **Flags:** S, Z, P, and AC are affected by the results of the operation. CY is not modified.

INX Rp

> **Description:** The contents of the register pair are incremented by 1 and stored in the register pair. The instruction views the two registers as a 16-bit number.
>
> **Bytes/M-Cycles/T-States:** 1/1/6
>
> Register Pair
>
> **Hex Codes:** 03   BC
> 13   DE
> 23   HL
> 33   SP
>
> **Flags:** No flags are affected.

JMP address(16b)

> **Description:** The program execution is transferred to the memory address specified.
>
> **Bytes/M-Cycles/T-States:** 3/3/10
>
> **Hex Codes:** C3
>
> **Flags:** No flags are affected.

JC address(16b)

> **Description:** Program execution is transferred to the memory address specified if the carry flag is set, or CY = 1. If CY = 0, no transfer takes place.

Bytes/M-Cycles/T-States: 3/2/7   if condition is not true

3/3/10   if condition is true

Hex Codes: DA

Flags: No flags are affected.

JNC address(16b)

Description: Program execution is transferred to the memory address specified if the carry flag is not set, or CY = 0. If CY = 1, no transfer takes place.

Bytes/M-Cycles/T-States: 3/2/7   if condition is not true

3/3/10   if condition is true

Hex Codes: D2

Flags: No flags are affected.

JP address(16b)

Description: Program execution is transferred to the memory address specified if positive, or S = 0. If S = 1, no transfer takes place.

Bytes/M-Cycles/T-States: 3/2/7   if condition is not true

3/3/10   if condition is true

Hex Codes: F2

Flags: No flags are affected.

JM address(16b)

Description: Program execution is transferred to the memory address specified if minus, or S = 1. If S = 0, no transfer takes place.

Bytes/M-Cycles/T-States: 3/2/7   if condition is not true

3/3/10   if condition is true

Hex Codes: FA

Flags: No flags are affected.

JPE address(16b)

Description: Program execution is transferred to the memory address specified if parity is even, or P = 1. If P = 0, no transfer takes place.

Bytes/M-Cycles/T-States: 3/2/7   if condition is not true

3/3/10   if condition is true

Hex Codes: EA

Flags: No flags are affected.

JPO address(16b)

Description: Program execution is transferred to the memory address specified if parity is odd, or P = 0. If P = 1, no transfer takes place.

Bytes/M-Cycles/T-States: 3/2/7   if condition is not true

3/3/10   if condition is true

Hex Codes: E2

Flags: No flags are affected.

JZ address(16b)

> **Description:** Program execution is transferred to the memory address specified if zero, or Z = 1. If Z = 0, no transfer takes place.
>
> **Bytes/M-Cycles/T-States:**   3/2/7   if condition is not true
>
>   3/3/10   if condition is true
>
> **Hex Codes:** CA
>
> **Flags:** No flags are affected.

JNZ address(16b)

> **Description:** Program execution is transferred to the memory address specified if not zero, or Z = 0. If Z = 1, no transfer takes place.
>
> **Bytes/M-Cycles/T-States:**   3/2/7   if condition is not true
>
>   3/3/10   if condition is true
>
> **Hex Codes:** C2
>
> **Flags:** No flags are affected.

LDA address(16b)

> **Description:** The contents of the memory location specified are transferred to the accumulator.
>
> **Bytes/M-Cycles/T-States:** 3/4/13
>
> **Hex Codes:** 3A
>
> **Flags:** No flags are affected.

LDAX Rp

> **Description:** The contents of the memory location pointed to by the register pair are loaded into the accumulator.
>
> **Bytes/M-Cycles/T-States:** 1/2/7
>
>   Register Pairs
>
> **Hex Codes:** 0A   BC
>
>   1A   DE
>
> **Flags:** No flags are affected.

LHLD address(16b)

> **Description:** The contents of the memory location specified are loaded into register L and the contents of the next memory location are loaded into register H.
>
> **Bytes/M-Cycles/T-States:** 3/5/16
>
> **Hex Codes:** 2A
>
> **Flags:** No flags are affected.

LXI Rp, data(16b)

> **Description:** The 16-bit data is loaded into the register pair.
>
> **Bytes/M-Cycles/T-States:** 3/3/10

Register Pair

Hex Codes: 01   BC

11   DE

21   HL

31   SP

Flags: No flags are affected.

MOV Rd, Rs

Description: The contents of the source register Rs are transferred into the destination register Rd.

Bytes/M-Cycles/T-States: 1/1/4

Hex Codes: Source Register

		A	B	C	D	E	H	L
		A	B	C	D	E	H	L
	A	7F	78	79	7A	7B	7C	7D
Destination	B	47	40	41	42	43	44	45
Register	C	41	48	49	4A	4B	4C	4D
	D	57	50	51	52	53	54	55
	E	5F	58	59	5A	5B	5C	5D
	H	67	60	61	62	63	64	65
	L	6F	68	69	6A	6B	6C	6D

Flags: No flags are affected.

MOV M, Rs

Description: The contents of the source register Rs are transferred to the memory location pointed to by HL.

Bytes/M-Cycles/T-States: 1/2/7

Hex Codes: Source Register

77   A

70   B

71   C

72   D

73   E

74   H

75   L

Flags: No flags are affected.

MOV Rd, M

Description: The contents of the memory location pointed to by HL are transferred to the destination register.

Bytes/M-Cycles/T-States: 1/2/7

Hex Codes: Destination Register

7E   A

46   B

4E   C

56	D
5E	E
66	H
6E	L

Flags: No flags are affected.

MVI R, data(8b)

Description: The 8 bits of data are stored in the register.

Bytes/M-Cycles/T-States: 2/2/7

Register

Hex Codes:
3E	A
06	B
0E	C
16	D
1E	E
26	H
2E	L

Flags: No flags are affected.

MVI M, data(8b)

Description: The 8 bits of data are stored in the memory location pointed to by HL.

Bytes/M-Cycles/T-States: 2/3/10

Hex Codes: 36

Flags: No flags are affected.

NOP

Description: No operation is performed. The instruction is fetched and decoded, but no operation is executed.

Bytes/M-Cycles/T-States: 1/1/4

Hex Codes: 00

Flags: No flags are affected.

ORA R

Description: The contents of the accumulator are logically OR'd with the contents of the register. The result is stored in the accumulator.

Bytes/M-Cycles/T-States: 1/1/4

Register

Hex Codes:
B7	A
B0	B
B1	C
B2	D
B3	E
B4	H
B5	L

Flags: Z, S, and P are affected based upon the operation. AC and CY are reset.

ORA M

> **Description:** The contents of the accumulator are logically OR'd with the contents of the memory location pointed to by HL.
> **Bytes/M-Cycles/T-States:** 1/2/7
> **Hex Codes:** B6
> **Flags:** Z, S, and P are affected based upon the operation. AC and CY are reset.

ORI data(8b)

> **Description:** The contents of the accumulator are logically OR'd with the 8 bits of data. The result is stored in the accumulator.
> **Bytes/M-Cycles/T-States:** 2/2/7
> **Hex Codes:** F6
> **Flags:** Z, S, and P are affected based upon the operation. AC and CY are reset.

OUT port address(8b)

> **Description:** The contents of the accumulator are copied out to the output port specified.
> **Bytes/M-Cycles/T-States:** 2/3/10
> **Hex Codes:** D3
> **Flags:** No flags are affected.

PCHL

> **Description:** The contents of registers H and L are copied into the program counter. H is the high-order bits and L is the low-order bits.
> **Bytes/M-Cycles/T-States:** 1/1/6
> **Hex Codes:** E9
> **Flags:** No flags are affected.

POP Rp

> **Description:** The contents of the memory location (stack) pointed to by the stack pointer are copied to the low-order register of the register pair. (C, E, L, and flags.) The stack pointer is then incremented and the contents of that memory location being pointed to are copied to the high-order register of the register pair.
> **Bytes/M-Cycles/T-States:** 1/3/10
>
> | | Register |
> | ------------ | -------- |
> | **Hex Codes:** C1 | BC |
> | D1 | DE |
> | E1 | HL |
> | F1 | PSW |
>
> **Flags:** No flags are affected.

PUSH Rp

> **Description:** The contents of the register pair are copied into the stack. The high-order register (B, D, H, A) is put on the stack first, then the contents of the low-order register (C, E, L, flags) are put onto the stack.

Bytes/M-Cycles/T-States: 1/3/12

Register Pair

Hex Codes: C5   B

           D5   D

           E5   H

           F5   PSW

Flags: No flags are affected.

RAL

Description: The contents of the accumulator are rotated left by one position through the carry flag.

Bytes/M-Cycles/T-States: 1/1/4

Hex Codes: 17

Flags: CY is modified according to bit D7. S, Z, P, and AC are not affected.

RAR

Description: The contents of the accumulator are rotated right by one position through the carry flag.

Bytes/M-Cycles/T-States: 1/1/4

Hex Codes: 1F

Flags: CY is modified according to bit D0. S, Z, P, and AC are not affected.

RLC

Description: The contents of the accumulator are rotated left one position. Bit D7 is placed in both D0 and CY.

Bytes/M-Cycles/T-States: 1/1/4

Hex Codes: 07

Flags: CY is modified according to bit D7. S, Z, P, and AC are not affected.

RRC

Description: The contents of the accumulator are rotated right by 1 bit. Bit D0 is placed in both D7 and CY at the same time.

Bytes/M-Cycles/T-States: 1/1/4

Hex Codes: 0F

Flags: CY is modified according to bit D0. S, Z, P, and AC are not affected.

RET

Description: The program sequence is transferred from the subroutine to the calling program. The two bytes from the top of the stack are copied into the Program Counter and this is the address that program execution begins.

Bytes/M-Cycles/T-States: 1/3/10

Hex Codes: C9

Flags: No flags are affected.

RC

Description: The program sequence is transferred from the subroutine to the calling program if the carry flag is set, or CY = 1. If CY = 0, no transfer takes place.

Bytes/M-Cycles/T-States:	1/1/6	if condition is not true
	1/3/12	if condition is true

Hex Codes: D8

Flags: No flags are affected.

RNC

Description: The program sequence is transferred from the subroutine to the calling program if the carry flag is not set, or CY = 0. If CY = 1, no transfer takes place.

Bytes/M-Cycles/T-States:	1/1/6	if condition is not true
	1/3/12	if condition is true

Hex Codes: D0

Flags: No flags are affected.

RP

Description: the program sequence is transferred from the subroutine to the calling program if positive, or S = 0. If S = 1, no transfer takes place.

Bytes/M-Cycles/T-States:	1/1/6	if condition is not true
	1/3/12	if condition is true

Hex Codes: F0

Flags: No flags are affected.

RM

Description: The program sequence is transferred from the subroutine to the calling program if minus, or S = 1. If S = 0, no transfer takes place.

Bytes/M-Cycles/T-States:	1/1/6	if condition is not true
	1/3/12	if condition is true

Hex Codes: F8

Flags: No flags are affected.

RPE

Description: The program sequence is transferred from the subroutine to the calling program if the parity is even, or P = 1. If P = 0, no transfer takes place.

Bytes/M-Cycles/T-States:	1/1/6	if condition is not true
	1/3/12	if condition is true

Hex Codes: E8

Flags: No flags are affected.

RPO

Description: The program sequence is transferred from the subroutine to the calling program if the parity is odd, or P = 0. If P = 1, no transfer takes place.

Bytes/M-Cycles/T-States: 1/1/6   if condition is not true

1/3/12   if condition is true

Hex Codes: E0

Flags: No flags are affected

RZ

Description: The program sequence is transferred from the subroutine to the calling program if zero, or Z = 1. If Z = 0, no transfer takes place.

Bytes/M-Cycles/T-States: 1/1/6   if condition is not true

1/3/12   if condition is true

Hex Codes: C8

Flags: No flags are affected.

RNZ

Description: The program sequence is transferred from the subroutine to the calling program if not zero, or Z = 0. If Z = 1, no transfer takes place.

Bytes/M-Cycles/T-States: 1/1/6   if condition is not true

1/3/12   if condition is true

Hex Codes: C0

Flags: No flags are affected.

RIM

Description: This instruction is used to both read in the status of interrupts 7.5, 6.5, and 5.5, as well as to read in the serial input data bit. An 8-bit word is read in and stored in the accumulator. The layout of that word is shown in Figure A1-1.

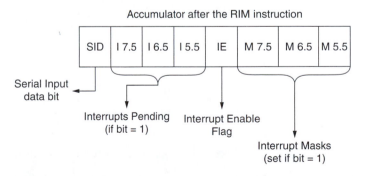

Accumulator after the RIM instruction

| SID | I 7.5 | I 6.5 | I 5.5 | IE | M 7.5 | M 6.5 | M 5.5 |

Serial Input data bit

Interrupts Pending (if bit = 1)   Interrupt Enable Flag

Interrupt Masks (set if bit = 1)

Figure A1-1 ■ The RIM instruction layout in the accumulator

Bytes/M-Cycles/T-States: 1/1/4

Hex Codes: 20

Flags: No flags are affected.

RST n (where n = 0 – 7)

Description: This instruction operates like a call instruction that goes to one of eight predetermined memory locations on page 0. Each instruction (RST 0–RST 7) goes to a specific address listed next.

Instruction	Restart Address
RST 0	0000
RST 1	0008
RST 2	0010
RST 3	0018
RST 4	0020
RST 5	0028
RST 6	0030
RST 7	0038

Bytes/M-Cycles/T-States: 1/3/12

Hex Codes:

C7	RST 0
CF	RST 1
D7	RST 2
DF	RST 3
E7	RST 4
EF	RST 5
F7	RST 6
FF	RST 7

Flags: No flags are accepted.

SBB R

Description: The contents of the register and the borrow flag are subtracted from the contents of the accumulator and the results are stored in the accumulator.

Bytes/M-Cycles/T-States: 1/1/4

Hex Codes:

	Register
9F	A
98	B
99	C
9A	D
9B	E
9C	H
9D	L

Flags: All flags are affected based upon the results of the operation.

SBB M

Description: The contents of the memory location pointed to by HL and the borrow flag are subtracted from the contents of the accumulator. The results are then stored in the accumulator.

Bytes/M-Cycles/T-States: 1/2/7

Hex Codes: 9E

Flags: All flags are affected based upon the results of the operation.

SBI data(8b)

Description: The 8 bits of data and the borrow flag are subtracted from the accumulator and the results are stored in the accumulator.

Bytes/M-Cycles/T-States: 2/2/7

Hex Codes: DE

Flags: All flags are affected based upon the results of the operation.

SHLD address(16b)

Description: The contents of register L are stored at the memory location specified and the contents of register H are stored at the next memory location by incrementing the operand by 1.

Bytes/M-Cycles/T-States: 3/5/16

Hex Codes: 22

Flags: No flags are affected.

SIM

Description: This is an instruction that is used to set the interrupt masks as well as set the serial output data bit. The accumulator is laid out as shown in Figure A1-2.

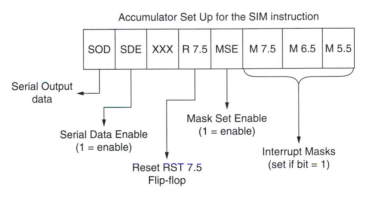

Figure A1-2 ■ The SIM instruction accumulator layout

Bytes/M-Cycles/T-States: 1/1/4

Hex Codes: 30

Flags: No flags are affected.

SPHL

Description: The contents of registers H and L are loaded into the Stack Pointer. H has the high-order part of the address, while L has the low-order portion.

Bytes/M-Cycles/T-States: 1/1/6

Hex Codes: F9

Flags: No flags are affected.

STA address(16b)

> **Description:** The contents of the accumulator are stored at the memory location specified.
>
> **Bytes/M-Cycles/T-States:** 3/4/13
>
> **Hex Codes:** 32
>
> **Flags:** No flags are affected.

STAX Rp (only BC or DE)

> **Description:** The contents of the accumulator are stored at the memory location pointed to by the register pair. The contents of the accumulator are not affected.
>
> **Bytes/M-Cycles/T-States:** 1/2/7
>
> Register Pair
>
> **Hex Codes:** 02   BC
>
>   12   DE
>
> **Flags:** No flags are affected.

STC

> **Description:** The carry flag, CY, is set to 1.
>
> **Bytes/M-Cycles/T-States:** 1/1/4
>
> **Hex Codes:** 37
>
> **Flags:** Only the CY flag is affected.

SUB R

> **Description:** The contents of the register are subtracted from the accumulator and the result is stored in the accumulator.
>
> **Bytes/M-Cycles/T-States:** 1/1/4
>
> Register
>
> **Hex Codes:** 97   A
>
>   90   B
>
>   91   C
>
>   92   D
>
>   93   E
>
>   94   H
>
>   95   L
>
> **Flags:** All flags are affected by the result of the operation.

SUB M

> **Description:** The contents of the memory location pointed to by HL are subtracted from the accumulator and the result is stored in the accumulator.
>
> **Bytes/M-Cycles/T-States:** 1/2/7
>
> **Hex Codes:** 96
>
> **Flags:** All flags are affected by the result of the operation.

SUI data(8b)

> **Description:** The 8-bit data is subtracted from the contents of the accumulator. The result is stored in the accumulator.
>
> **Bytes/M-Cycles/T-States:** 2/2/7
>
> **Hex Codes:** D6
>
> **Flags:** All flags are affected by the result of the operation.

XCHG

> **Description:** The contents of register H and register D are exchanged, and the contents of register L and register E are exchanged.
>
> **Bytes/M-Cycles/T-States:** 1/1/4
>
> **Hex Codes:** EB
>
> **Flags:** No flags are affected.

XRA R

> **Description:** The contents of the register are Exclusive OR'd with the contents of the accumulator. The results are then stored in the accumulator.
>
> **Bytes/M-Cycles/T-States:** 1/1/4

	Register
Hex Codes: AF	A
A8	B
A9	C
AA	D
AB	E
AC	H
AD	L

> **Flags:** Z, S, and P are affected based upon the operation. CY and AC are reset.

XRA M

> **Description:** The contents of the memory location pointed to by HL are Exclusive OR'd with the contents of the accumulator. The results are stored in the accumulator.
>
> **Bytes/M-Cycles/T-States:** 1/2/7
>
> **Hex Codes:** AE
>
> **Flags:** Z, S, and P are affected based upon the operation. CY and AC are reset.

XRI data(8b)

> **Description:** The 8 bits of data are Exclusive OR'd with the contents of the accumulator. The results are then stored in the accumulator.
>
> **Bytes/M-Cycles/T-States:** 2/2/7
>
> **Hex Codes:** EE
>
> **Flags:** Z, S, and P are affected based upon the operation. CY and AC are reset.

XTHL

Description: The contents of the L register are exchanged with the stack location pointed to by the stack pointer. The contents of the H register are exchanged with the stack location pointed to by the stack pointer +1. The stack pointer remains unchanged.

Bytes/M-Cycles/T-States: 1/5/16

Hex Codes: E3

Flags: No flags are affected.

Appendix II—The 8085 Microprocessor Specification Sheet

intel.

8085AH/8085AH-2/8085AH-1
8-BIT HMOS MICROPROCESSORS

- Single +5V Power Supply with 10% Voltage Margins
- 3 MHz, 5 MHz and 6 MHz Selections Available
- 20% Lower Power Consumption than 8085A for 3 MHz and 5 MHz
- 1.3 μs Instruction Cycle (8085AH); 0.8 μs (8085AH-2); 0.67 μs (8085AH-1)
- 100% Software Compatible with 8080A
- On-Chip Clock Generator (with External Crystal, LC or RC Network)

- On-Chip System Controller; Advanced Cycle Status Information Available for Large System Control
- Four Vectored Interrupt Inputs (One Is Non-Maskable) Plus an 8080A-Compatible Interrupt
- Serial In/Serial Out Port
- Decimal, Binary and Double Precision Arithmetic
- Direct Addressing Capability to 64K Bytes of Memory
- Available in 40-Lead Cerdip and Plastic Packages
  (See Packaging Spec., Order #231369)

The Intel 8085AH is a complete 8-bit parallel Central Processing Unit (CPU) implemented in N-channel, depletion load, silicon gate technology (HMOS). Its instruction set is 100% software compatible with the 8080A microprocessor, and it is designed to improve the present 8080A's performance by higher system speed. Its high level of system integration allows a minimum system of three IC's [8085AH (CPU), 8156H (RAM/IO) and 8755A (EPROM/IO)] while maintaining total system expandability. The 8085AH-2 and 8085AH-1 are faster versions of the 8085AH.

The 8085AH incorporates all of the features that the 8224 (clock generator) and 8228 (system controller) provided for the 8080A, thereby offering a higher level of system integration.

The 8085AH uses a multiplexed data bus. The address is split between the 8-bit address bus and the 8-bit data bus. The on-chip address latches of 8155H/8156H/8755A memory products allow a direct interface with the 8085AH.

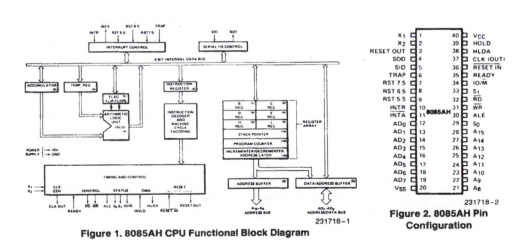

Figure 1. 8085AH CPU Functional Block Diagram

231718-1

231718-2

Figure 2. 8085AH Pin Configuration

September 1987
Order Number: 231718-001

8085AH/8085AH-2/8085AH-1   **intel**®

Table 1. Pin Description

Symbol	Type	Name and Function
A$_8$–A$_{15}$	O	**ADDRESS BUS:** The most significant 8 bits of memory address or the 8 bits of the I/O address, 3-stated during Hold and Halt modes and during RESET.
AD$_{0-7}$	I/O	**MULTIPLEXED ADDRESS/DATA BUS:** Lower 8 bits of the memory address (or I/O address) appear on the bus during the first clock cycle (T state) of a machine cycle. It then becomes the data bus during the second and third clock cycles.
ALE	O	**ADDRESS LATCH ENABLE:** It occurs during the first clock state of a machine cycle and enables the address to get latched into the on-chip latch of peripherals. The falling edge of ALE is set to guarantee setup and hold times for the address information. The falling edge of ALE can also be used to strobe the status information. ALE is never 3-stated.
S$_0$, S$_1$ and IO/$\overline{\text{M}}$	O	**MACHINE CYCLE STATUS:** IO/$\overline{\text{M}}$ S$_1$ S$_0$ Status 0 0 1 Memory write 0 1 0 Memory read 1 0 1 I/O write 1 1 0 I/O read 0 1 1 Opcode fetch 1 1 1 Interrupt Acknowledge * 0 0 Halt * X X Hold * X X Reset * = 3-state (high impedance) X = unspecified S$_1$ can be used as an advanced R/$\overline{\text{W}}$ status. IO/$\overline{\text{M}}$, S$_0$ and S$_1$ become valid at the beginning of a machine cycle and remain stable throughout the cycle. The falling edge of ALE may be used to latch the state of these lines.
$\overline{\text{RD}}$	O	**READ CONTROL:** A low level on $\overline{\text{RD}}$ indicates the selected memory or I/O device is to be read and that the Data Bus is available for the data transfer, 3-stated during Hold and Halt modes and during RESET.
$\overline{\text{WR}}$	O	**WROTE CONTROL:** A low level on $\overline{\text{WR}}$ indicates the data on the Data Bus is to be written into the selected memory or I/O location. Data is set up at the trailing edge of $\overline{\text{WR}}$. 3-stated during Hold and Halt modes and during RESET.
READY	I	**READY:** If READY is high during a read or write cycle, it indicates that the memory or peripheral is ready to send or receive data. If READY is low, the CPU will wait an integral number of clock cycles for READY to go high before completing the read or write cycle. READY must conform to specified setup and hold times.
HOLD	I	**HOLD:** Indicates that another master is requesting the use of the address and data buses. The CPU, upon receiving the hold request, will relinquish the use of the bus as soon as the completion of the current bus transfer. Internal processing can continue. The processor can regain the bus only after the HOLD is removed. When the HOLD is acknowledged, the Address, Data $\overline{\text{RD}}$, $\overline{\text{WR}}$, and IO/$\overline{\text{M}}$ lines are 3-stated.
HLDA	O	**HOLD ACKNOWLEDGE:** Indicates that the CPU has received the HOLD request and that it will relinquish the bus in the next clock cycle. HILDA goes low after the Hold request is removed. The CPU takes the bus one half clock cycle after HLDA goes low.
INTR	I	**INTERRUPT REQUEST:** Is used as a general purpose interrupt. It is sampled only during the next to the last clock cycle of an instruction and during Hold and Halt states. If it is active, the Program Counter (PC) will be inhibited from incrementing and an $\overline{\text{INTA}}$ will be issued. During this cycle a RESTART or CALL instruction can be inserted to jump to the interrupt service routine. The INTR is enabled and disabled by software. It is disabled by Reset and immediately after an interrupt is accepted.

intel.

Table 1. Pin Description (Continued)

Symbol	Type	Name and Function
INTA	O	**INTERRUPT ACKNOWLEDGE:** Is used instead of (and has the same timing as) RD during the Instruction cycle after an INTR is accepted. It can be used to activate an 8259A Interrupt chip or some other interrupt port.
RST 5.5 RST 6.5 RST 7.5	I	**RESTART INTERRUPTS:** These three inputs have the same timing as INTR except they cause an internal RESTART to be automatically inserted. The priority of these interrupt is ordered as shown in Table 2. These interrupts have a higher priority than INTR. In addition, they may be individually masked out using the SIM instruction.
TRAP	I	**TRAP:** Trap interrupt is a non-maskable RESTART interrupt. It is recognized at the same time as INTR or RST 5.5–7.5. It is unaffected by any mask or Interrupt Enable. It has the highest priority of any interrupt. (See Table 2.)
RESET IN	I	**RESET IN:** Sets the Program Counter to zero and resets the Interrupt Enable and HLDA flip-flops. The data and address buses and the control lines are 3-stated during RESET and because of the asynchronous nature of RESET, the processor's internal registers and flags may be altered by RESET with unpredictable results. RESET IN is a Schmitt-triggered input, allowing connection to an R-C network for power-on RESET delay (see Figure 3). Upon power-up, RESET IN must remain low for at least 10 ms after minimum V$_{CC}$ has been reached. For proper reset operation after the power-up duration, RESET IN should be kept low a minimum of three clock periods. The CPU is held in the reset condition as long as RESET IN is applied.
RESET OUT	O	**RESET OUT:** Reset Out indicates CPU is being reset. Can be used as a system reset. The signal is synchronized to the processor clock and lasts an integral number of clock periods.
X$_1$, X$_2$	I	**X$_1$ and X$_2$:** Are connected to a crystal, LC, or RC network to drive the internal clock generator. X$_1$ can also be an external clock input from a logic gate. The input frequency is divided by 2 to give the processor's internal operating frequency.
CLK	O	**CLOCK:** Clock output for use as a system clock. The period of CLK is twice the X$_1$, X$_2$ input period.
SID	I	**SERIAL INPUT DATA LINE:** The data on this line is loaded into accumulator bit 7 whenever a RIM instruction is executed.
SOD	O	**SERIAL OUTPUT DATA LINE:** The output SOD is set or reset as specified by the SIM instruction.
V$_{CC}$		**POWER:** + 5 volt supply.
V$_{SS}$		**GROUND:** Reference.

Table 2. Interrupt Priority, Restart Address and Sensitivity

Name	Priority	Address Branched to[1] When Interrupt Occurs	Type Trigger
TRAP	1	24H	Rising Edge AND High Level until Sampled
RST 7.5	2	3CH	Rising Edge (Latched)
RST 6.5	3	34H	High Level until Sampled
RST 5.5	4	2CH	High Level until Sampled
INTR	5	(Note 2)	High Level until Sampled

NOTES:
1. The processor pushes the PC on the stack before branching to the indicated address.
2. The address branched to depends on the instruction provided to the CPU when the interrupt is acknowledged.

8085AH/8085AH-2/8085AH-1

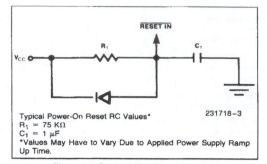

Typical Power-On Reset RC Values*
$R_1 = 75$ KΩ
$C_1 = 1$ μF
*Values May Have to Vary Due to Applied Power Supply Ramp Up Time.

231718–3

Figure 3. Power-On Reset Circuit

FUNCTIONAL DESCRIPTION

The 8085AH is a complete 8-bit parallel central processor. It is designed with N-channel, depletion load, silicon gate technology (HMOS), and requires a single +5V supply. Its basic clock speed is 3 MHz (8085AH), 5 MHz (8085AH-2), or 6 MHz (8085-AH-1), thus improving on the present 8080A's performance with higher system speed. Also it is designed to fit into a minimum system of three IC's: The CPU (8085AH), a RAM/IO (8156H), and an EPROM/IO chip (8755A).

The 8085AH has twelve addressable 8-bit registers. Four of them can function only as two 16-bit register pairs. Six others can be used interchangeably as 8-bit registers or as 16-bit register pairs. The 8085AH register set is as follows:

Mnemonic	Register	Contents
ACC or A	Accumulator	8 Bits
PC	Program Counter	16-Bit Address
BC, DE, HL	General-Purpose Registers; data pointer (HL)	8-Bits x 6 or 16 Bits x 3
SP	Stack Pointer	16-Bit Address
Flags or F	Flag Register	5 Flags (8-Bit Space)

The 8085AH uses a multiplexed Data Bus. The address is split between the higher 8-bit Address Bus and the lower 8-bit Address/Data Bus. During the first T state (clock cycle) of a machine cycle the low order address is sent out on the Address/Data bus. These lower 8 bits may be latched externally by the Address Latch Enable signal (ALE). During the rest of the machine cycle the data bus is used for memory or I/O data.

The 8085AH provides $\overline{RD}$, $\overline{WR}$, S_0, S_1, and IO/$\overline{M}$ signals for bus control. An Interrupt Acknowledge signal ($\overline{INTA}$) is also provided. HOLD and all Interrupts are synchronized with the processor's internal clock. The 8085AH also provides Serial Input Data

(SID) and Serial Output Data (SOD) lines for simple serial interface.

In addition to these features, the 8085AH has three maskable, vector interrupt pins, one nonmaskable TRAP interrupt, and a bus vectored interrupt, INTR.

INTERRUPT AND SERIAL I/O

The 8085AH has 5 interrupt inputs: INTR, RST 5.5, RST 6.5, RST 7.5, and TRAP. INTR is identical in function to the 8080A INT. Each of the three RESTART inputs, 5.5, 6.5, and 7.5, has a programmable mask. TRAP is also a RESTART interrupt but it is nonmaskable.

The three maskable interrupt cause the internal execution of RESTART (saving the program counter in the stack and branching to the RESTART address) if the interrupts are enabled and if the interrupt mask is not set. The nonmaskable TRAP causes the internal execution of a RESTART vector independent of the state of the interrupt enable or masks. (See Table 2.)

There are two different types of inputs in the restart interrupts. RST 5.5 and RST 6.5 are *high level-sensitive* like INTR (and INT on the 8080) and are recognized with the same timing as INTR. RST 7.5 is *rising edge-sensitive.*

For RST 7.5, only a pulse is required to set an internal flip-flop which generates the internal interrupt request (a normally high level signal with a low going pulse is recommended for highest system noise immunity). The RST 7.5 request flip-flop remains set until the request is serviced. Then it is reset automatically. This flip-flop may also be reset by using the SIM instruction or by issuing a $\overline{RESET IN}$ to the 8085AH. The RST 7.5 internal flip-flop will be set by a pulse on the RST 7.5 pin even when the RST 7.5 interrupt is masked out.

The status of the three RST interrupt masks can only be affected by the SIM instruction and $\overline{RESET IN}$. (See SIM, Chapter 5 of the 8080/8085 User's Manual.)

The interrupts are arranged in a fixed priority that determines which interrupt is to be recognized if more than one is pending as follows: TRAP—highest priority, RST 7.5, RST 6.5, RST 5.5, INTR—lowest priority. This priority scheme does not take into account the priority of a routine that was started by a higher priority interrupt. RST 5.5 can interrupt an RST 7.5 routine if the interrupts are re-enabled before the end of the RST 7.5 routine.

The TRAP interrupt is useful for catastrophic events such as power failure or bus error. The TRAP input is recognized just as any other interrupt but has the

intel.

highest priority. It is not affected by any flag or mask. The TRAP input is both *edge and level sensitive.* The TRAP input must go high and remain high until it is acknowledged. It will not be recognized again until it goes low, then high again. This avoids any false triggering due to noise or logic glitches. Figure 4 illustrates the TRAP interrupt request circuitry within the 8085AH. Note that the servicing of any interrupt (TRAP, RST 7.5, RST 6.5, RST 5.5, INTR) disables all future interrupts (except TRAPs) until an EI instruction is executed.

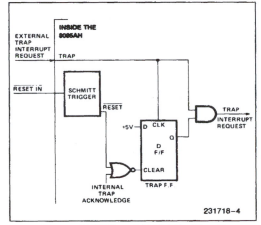

Figure 4. TRAP and RESET In Circuit

The TRAP interrupt is special in that it disables interrupts, but preserves the previous interrupt enable status. Performing the first RIM instruction following a TRAP interrupt allows you to determine whether interrupts were enabled or disabled prior to the TRAP. All subsequent RIM instructions provide current interrupt enable status. Performing a RIM instruction following INTR, or RST 5.5–7.5 will provide current Interrupt Enable status, revealing that interrupts are disabled. See the description of the RIM instruction in the 8080/8085 Family User's Manual.

The serial I/O system is also controlled by the RIM and SIM instruction. SID is read by RIM, and SIM sets the SOD data.

DRIVING THE X_1 AND X_2 INPUTS

You may drive the clock inputs of the 8085AH, 8085AH-2, or 8085AH-1 with a crystal, an LC tuned circuit, an RC network, or an external clock source. The crystal frequency must be at least 1 MHz, and must be twice the desired internal clock frequency;

hence, the 8085AH is operated with a 6 MHz crystal (for 3 MHz clock), the 8085AH-2 operated with a 10 MHz crystal (for 5 MHz clock), and the 8085AH-1 can be operated with a 12 MHz crystal (for 6 MHz clock). If a crystal is used, it must have the following characteristics:

Parallel resonance at twice the clock frequency desired
C_L (load capacitance) $\leq$ 30 pF
C_S (Shunt capacitance) $\leq$ 7 pF
R_S (equivalent shunt resistance) $\leq$ 75Ω
Drive level: 10 mW
Frequency tolerance: $\pm$0.005% (suggested)

Note the use of the 20 pF capacitor between X_2 and ground. This capacitor is required with crystal frequencies below 4 MHz to assure oscillator startup at the correct frequency. A parallel-resonant LC citcuit may be used as the frequency-determining network for the 8085AH, providing that its frequency tolerance of approximately $\pm$10% is acceptable. The components are chosen from the formula:

$$f = \frac{1}{2\pi\sqrt{L(C_{ext} + C_{int})}}$$

To minimize variations in frequency, it is recommended that you choose a value for C_{ext} that is at least twice that of C_{int}, or 30 pF. The use of an LC circuit is not recommended for frequencies higher than approximately 5 MHz.

An RC circuit may be used as the frequency-determining network for the 8085AH if maintaining a precise clock frequency is of no importance. Variations in the on-chip timing generation can cause a wide variation in frequency when using the RC mode. Its advantage is its low component cost. The driving frequency generated by the circuit shown is approximately 3 MHz. It is not recommended that frequencies greatly higher or lower than this be attempted.

Figure 5 shows the recommended clock driver circuits. Note in d and e that pullup resistors are required to assure that the high level voltage of the input is at least 4V and maximum low level voltage of 0.8V.

For driving frequencies up to and including 6 MHz you may supply the driving signal to X_1 and leave X_2 open-circuited (Figure 5d). If the driving frequency is from 6 MHz to 12 MHz, stability of the clock generator will be improved by driving both X_1 and X_2 with a push-pull source (Figure 5e). To prevent self-oscillation of the 8085AH, be sure that X_2 is not coupled back to X_1 through the driving circuit.

8085AH/8085AH-2/8085AH-1   **intel**®

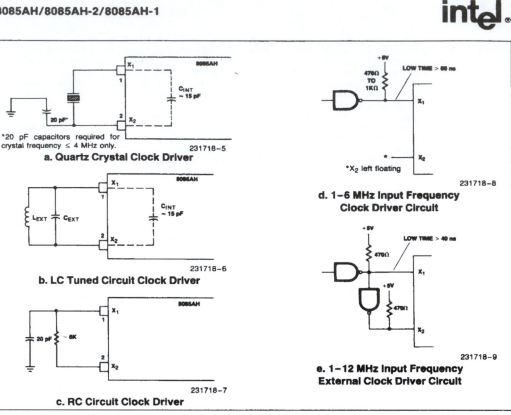

Figure 5. Clock Driver Circuits

GENERATING AN 8085AH WAIT STATE

If your system requirements are such that slow memories or peripheral devices are being used, the circuit shown in Figure 6 may be used to insert one WAIT state in each 8085AH machine cycle.

The D flip-flops should be chosen so that
• CLK is rising edge-triggered
• CLEAR is low-level active.

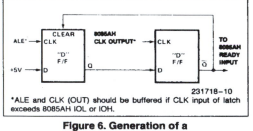

*ALE and CLK (OUT) should be buffered if CLK input of latch exceeds 8085AH IOL or IOH.

Figure 6. Generation of a Wait State for 8085AH CPU

As in the 8080, the READY line is used to extend the read and write pulse lengths so that the 8085AH can be used with slow memory. HOLD causes the CPU to relinquish the bus when it is through with it by floating the Address and Data Buses.

SYSTEM INTERFACE

The 8085AH family includes memory components, which are directly compatible to the 8085AH CPU. For example, a system consisting of the three chips, 8085AH, 8156H and 8755A will have the following features:

• 2K Bytes EPROM
• 256 Bytes RAM
• 1 Timer/Counter
• 4 8-bit I/O Ports
• 1 6-bit I/O Port
• 4 Interrupt Levels
• Serial In/Serial Out Ports

intel.

This minimum system, using the standard I/O technique is as shown in Figure 7.

In addition to the standard I/O, the memory mapped I/O offers an efficient I/O addressing technique. With this technique, an area of memory address space is assigned for I/O address, thereby, using the memory address for I/O manipulation. Figure 8

shows the system configuration of Memory Mapped I/O using 8085AH.

The 8085AH CPU can also interface with the standard memory that does *not* have the multiplexed address/data bus. It will require a simple 8-bit latch as shown in Figure 9.

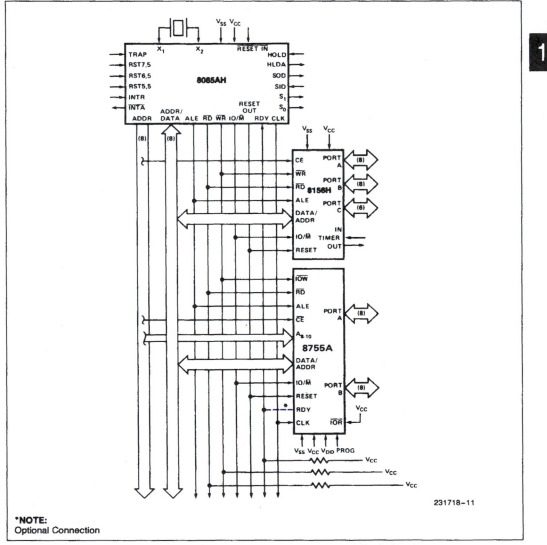

231718–11

*NOTE:
Optional Connection

Figure 7. 8085AH Minimum System (Standard I/O Technique)

8085AH/8085AH-2/8085AH-1

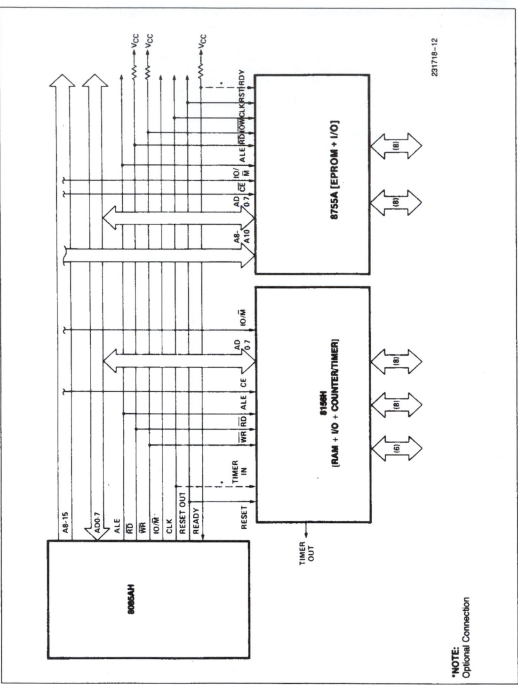

Figure 8. 8085 Minimum System (Memory Mapped I/O)

231718–12

*NOTE:
Optional Connection

intel® 8085AH/8085AH-2/8085AH-1

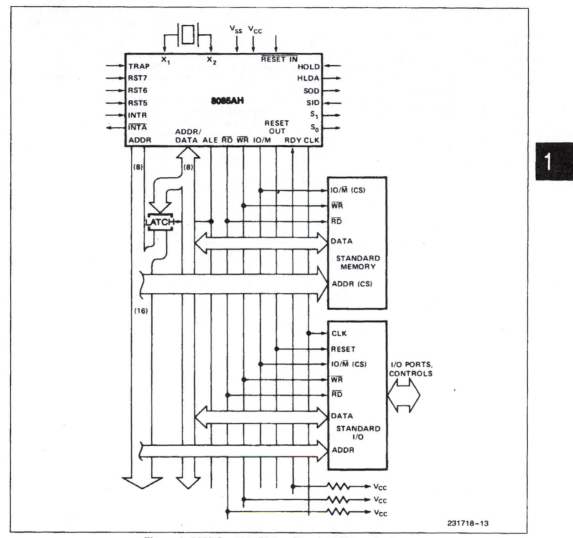

Figure 9. 8085 System (Using Standard Memories)

231718-13

8085AH/8085AH-2/8085AH-1

intel®

BASIC SYSTEM TIMING

The 8085AH has a multiplexed Data Bus. ALE is used as a strobe to sample the lower 8-bits of address on the Data Bus. Figure 10 shows an instruction fetch, memory read and I/O write cycle (as would occur during processing of the OUT instruction). Note that during the I/O write and read cycle that the I/O port address is copied on both the upper and lower half of the address.

There are seven possible types of machine cycles. Which of these seven takes place is defined by the status of the three status lines (IO/$\overline{M}$, S_1, S_0) and

the three control signals ($\overline{RD}$, $\overline{WR}$, and $\overline{INTA}$). (See Table 3.) The status lines can be used as advanced controls (for device selection, for example), since they become active at the T_1 state, at the outset of each machine cycle. Control lines $\overline{RD}$ and $\overline{WR}$ become active later, at the time when the transfer of data is to take place, so are used as command lines.

A machine cycle normally consists of three T states, with the exception of OPCODE FETCH, which normally has either four or six T states (unless WAIT or HOLD states are forced by the receipt of READY or HOLD inputs). Any T state must be one of ten possible states, shown in Table 4.

Table 3. 8085AH Machine Cycle Chart

Machine Cycle			Status			Control		
			IO/$\overline{M}$	S1	S0	$\overline{RD}$	$\overline{WR}$	$\overline{INTA}$
OPCODE FETCH	(OF)		0	1	1	0	1	1
MEMORY READ	(MR)		0	1	0	0	1	1
MEMORY WRITE	(MW)		0	0	1	1	0	1
I/O READ	(IOR)		1	1	0	0	1	1
I/O WRITE	(IOW)		1	0	1	1	0	1
ACKNOWLEDGE OF INTR	(INA)		1	1	1	1	1	0
BUS IDLE	(BI):	DAD	0	1	0	1	1	1
		ACK.OF RST,TRAP	1	1	1	1	1	1
		HALT	TS	0	0	TS	TS	1

Table 4. 8085AH Machine State Chart

Machine State	Status & Buses				Control		
	S1,S0	IO/$\overline{M}$	A_8–A_{15}	AD_0–AD_7	$\overline{RD}$, $\overline{WR}$	$\overline{INTA}$	ALE
T_1	X	X	X	X	1	1	1*
T_2	X	X	X	X	X	X	0
T_{WAIT}	X	X	X	X	X	X	0
T_3	X	X	X	X	X	X	0
T_4	1	0†	X	TS	1	1	0
T_5	1	0†	X	TS	1	1	0
T_6	1	0†	X	TS	1	1	0
T_{RESET}	X	TS	TS	TS	TS	1	0
T_{HALT}	0	TS	TS	TS	TS	1	0
T_{HOLD}	X	TS	TS	TS	TS	1	0

0 = Logic "0"   TS = High Impedance
1 = Logic "1"   X = Unspecified
*ALE not generated during 2nd and 3rd machine cycles of DAD instruction.
†IO/$\overline{M}$ = 1 during T_4–T_6 of INA machine cycle.

intel®

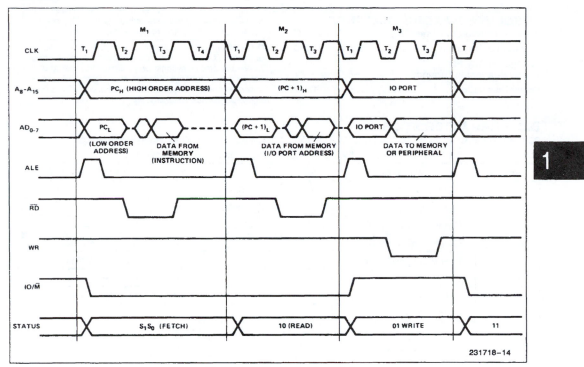

Figure 10. 8085AH Basic System Timing

231718-14

1

8085AH/8085AH-2/8085AH-1   **intel**®

ABSOLUTE MAXIMUM RATINGS*

Ambient Temperature under Bias ......0°C to 70°C

Storage Temperature ....−65°C to +150°C

Voltage on Any Pin
  with Respect to Ground..........−0.5V to +7V

Power Dissipation.........................1.5W

NOTICE: This is a production data sheet. The specifications are subject to change without notice.

*WARNING: Stressing the device beyond the "Absolute Maximum Ratings" may cause permanent damage. These are stress ratings only. Operation beyond the "Operating Conditions" is not recommended and extended exposure beyond the "Operating Conditions" may affect device reliability.

D.C. CHARACTERISTICS

8085AH, 8085AH-2: T_A = 0°C to 70°C, V_{CC} = 5V ±10%, V_{SS} = 0V; unless otherwise specified*

8085AH-1: T_A = 0°C to 70°C, V_{CC} = 5V ±5%, V_{SS} = 0V; unless otherwise specified*

Symbol	Parameter	Min	Max	Units	Test Conditions
V_{IL}	Input Low Voltage	−0.5	+0.8	V	
V_{IH}	Input High Voltage	2.0	V_{CC} +0.5	V	
V_{OL}	Output Low Voltage		0.45	V	I_{OL} = 2 mA
V_{OH}	Output High Voltage	2.4		V	I_{OH} = −400 µA
I_{CC}	Power Supply Current		135	mA	8085AH, 8085AH-2
			200	mA	8085AH-1
I_{IL}	Input Leakage		±10	µA	0 ≤ V_{IN} ≤ V_{CC}
I_{LO}	Output Leakage		±10	µA	0.45V ≤ V_{OUT} ≤ V_{CC}
V_{ILR}	Input Low Level, RESET	−0.5	+0.8	V	
V_{IHR}	Input High Level, RESET	2.4	V_{CC} + 0.5	V	
V_{HY}	Hysteresis, RESET	0.15		V	

A.C. CHARACTERISTICS

8085AH, 8085AH-2: T_A = 0°C to 70°C, V_{CC} = 5V ±10%, V_{SS} = 0V*

8085AH-1: T_A = 0°C to 70°C, V_{CC} = 5V ±5%, V_{SS} = 0V

Symbol	Parameter	8085AH [2]		8085AH-2 [2]		8085AH-1 [2]		Units
		Min	Max	Min	Max	Min	Max	
t_{CYC}	CLK Cycle Period	320	2000	200	2000	167	2000	ns
t_1	CLK Low Time (Standard CLK Loading)	80		40		20		ns
t_2	CLK High Time (Standard CLK Loading)	120		70		50		ns
t_r, t_f	CLK Rise and Fall Time		30		30		30	ns
t_{XKR}	X_1 Rising to CLK Rising	20	120	20	100	20	100	ns
t_{XKF}	X_1 Rising to CLK Falling	20	150	20	110	20	110	ns
t_{AC}	A_{8-15} Valid to Leading Edge of Control [1]	270		115		70		ns
t_{ACL}	A_{0-7} Valid to Leading Edge of Control	240		115		60		ns
t_{AD}	A_{0-15} Valid to Valid Data In		575		350		225	ns
t_{AFR}	Address Float after Leading Edge of READ (INTA)		0		0		0	ns
t_{AL}	A_{8-15} Valid before Trailing Edge of ALE [1]	115		50		25		ns

*NOTE:
For Extended Temperature EXPRESS use M8085AH Electricals Parameters.

intel.

A.C. CHARACTERISTICS (Continued)

Symbol	Parameter	8085AH (2)		8085AH-2 (2)		8085AH-1 (2)		Units
		Min	Max	Min	Max	Min	Max	
t_{ALL}	A_{0-7} Valid before Trailing Edge of ALE	90		50		25		ns
t_{ARY}	READY Valid from Address Valid		220		100		40	ns
t_{CA}	Address (A_{8-15}) Valid after Control	120		60		30		ns
t_{CC}	Width of Control Low ($\overline{RD}$, $\overline{WR}$, $\overline{INTA}$) Edge of ALE	400		230		150		ns
t_{CL}	Trailing Edge of Control to Leading Edge of ALE	50		25		0		ns
t_{DW}	Data Valid to Trialing Edge of $\overline{WRITE}$	420		230		140		ns
t_{HABE}	HLDA to Bus Enable		210		150		150	ns
t_{HABF}	Bus Float after HLDA		210		150		150	ns
t_{HACK}	HLDA Valid to Trailing Edge of CLK	110		40		0		ns
t_{HDH}	HOLD Hold Time	0		0		0		ns
t_{HDS}	HOLD Setup Time to Trailing Edge of CLK	170		120		120		ns
t_{INH}	INTR Hold Time	0		0		0		ns
t_{INS}	INTR, RST, and TRAP Setup Time to Falling Edge of CLK	160		150		150		ns
t_{LA}	Address Hold Time after ALE	100		50		20		ns
t_{LC}	Trailing Edge of ALE to Leading Edge of Control	130		60		25		ns
t_{LCK}	ALE Low During CLK High	100		50		15		ns
t_{LDR}	ALE to Valid Data during Read		460		270		175	ns
t_{LDW}	ALE to Valid Data during Write		200		140		110	ns
t_{LL}	ALE Width	140		80		50		ns
t_{LRY}	ALE to READY Stable		110		30		10	ns
t_{RAE}	Trailing Edge of $\overline{READ}$ to Re-Enabling of Address	150		90		50		ns
t_{RD}	$\overline{READ}$ (or $\overline{INTA}$) to Valid Data		300		150		75	ns
t_{RV}	Control Trailing Edge to Leading Edge of Next Control	400		220		160		ns
t_{RDH}	Data Hold Time after $\overline{READ}$ $\overline{INTA}$	0		0		0		ns
t_{RYH}	READY Hold Time	0		0		5		ns
t_{RYS}	READY Setup Time to Leading Edge of CLK	110		100		100		ns
t_{WD}	Data Valid after Trailing Edge of $\overline{WRITE}$	100		60		30		ns
t_{WDL}	LEADING Edge of $\overline{WRITE}$ to Data Valid		40		20		30	ns

NOTES:
1. A_8-A_{15} address Specs apply IO/$\overline{M}$, S_0, and S_1 except A_8-A_{15} are undefined during T_4-T_6 of OF cycle whereas IO/$\overline{M}$, S_0, and S_1 are stable.
2. *Test Conditions:* t_{CYC} = 320 ns (8085AH)/200 ns (8085AH-2);/167 ns (8085AH-1); C_L = 150 pF.
3. For all output timing where C ≠ 150 pF use the following correction factors:
   25 pF ≤ C_L < 150 pF: −0.10 ns/pF
   150 pF < C_L ≤ 300 pF: +0.30 ns/pF
4. Output timings are measured with purely capacitive load.
5. To calculate timing specifications at other values of t_{CYC} use Table 5.

8085AH/8085AH-2/8085AH-1

intel.

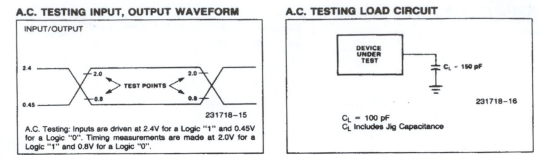

A.C. TESTING INPUT, OUTPUT WAVEFORM

INPUT/OUTPUT

2.4

2.0   TEST POINTS   2.0

0.8   0.8

0.45

231718–15

A.C. Testing: Inputs are driven at 2.4V for a Logic "1" and 0.45V for a Logic "0". Timing measurements are made at 2.0V for a Logic "1" and 0.8V for a Logic "0".

A.C. TESTING LOAD CIRCUIT

DEVICE UNDER TEST

C_L – 150 pF

231718–16

$C_L = 100$ pF
C_L Includes Jig Capacitance

Table 5. Bus Timing Specification as a T_{CYC} Dependent

Symbol	8085AH	8085AH-2	8085AH-1	
t_{AL}	$(1/2)T - 45$	$(1/2)T - 50$	$(1/2)T - 58$	Minimum
t_{LA}	$(1/2)T - 60$	$(1/2)T - 50$	$(1/2)T - 63$	Minimum
t_{LL}	$(1/2)T - 20$	$(1/2)T - 20$	$(1/2)T - 33$	Minimum
t_{LCK}	$(1/2)T - 60$	$(1/2)T - 50$	$(1/2)T - 68$	Minimum
t_{LC}	$(1/2)T - 30$	$(1/2)T - 40$	$(1/2)T - 58$	Minimum
t_{AD}	$(5/2 + N)T - 225$	$(5/2 + N)T - 150$	$(5/2 + N)T - 192$	Maximum
t_{RD}	$(3/2 + N)T - 180$	$(3/2 + N)T - 150$	$(3/2 + N)T - 175$	Maximum
t_{RAE}	$(1/2)T - 10$	$(1/2)T - 10$	$(1/2)T - 33$	Minimum
t_{CA}	$(1/2)T - 40$	$(1/2)T - 40$	$(1/2)T - 53$	Minimum
t_{DW}	$(3/2 + N)T - 60$	$(3/2 + N)T - 70$	$(3/2 + N)T - 110$	Minimum
t_{WD}	$(1/2)T - 60$	$(1/2)T - 40$	$(1/2)T - 53$	Minimum
t_{CC}	$(3/2 + N)T - 80$	$(3/2 + N)T - 70$	$(3/2 + N)T - 100$	Minimum
t_{CL}	$(1/2)T - 110$	$(1/2)T - 75$	$(1/2)T - 83$	Minimum
t_{ARY}	$(3/2)T - 260$	$(3/2)T - 200$	$(3/2)T - 210$	Maximum
t_{HACK}	$(1/2)T - 50$	$(1/2)T - 60$	$(1/2)T - 83$	Minimum
t_{HABF}	$(1/2)T + 50$	$(1/2)T + 50$	$(1/2)T + 67$	Maximum
t_{HABE}	$(1/2)T + 50$	$(1/2)T + 50$	$(1/2)T + 67$	Maximum
t_{AC}	$(2/2)T - 50$	$(2/2)T - 85$	$(2/2)T - 97$	Minimum
t_1	$(1/2)T - 80$	$(1/2)T - 60$	$(1/2)T - 63$	Minimum
t_2	$(1/2)T - 40$	$(1/2)T - 30$	$(1/2)T - 33$	Minimum
t_{RV}	$(3/2)T - 80$	$(3/2)T - 80$	$(3/2)T - 90$	Minimum
t_{LDR}	$(4/2 + N)T - 180$	$(4/2)T - 130$	$(4/2)T - 159$	Maximum

NOTE:
N is equal to the total WAIT states. $T = t_{CYC}$.

intel.

WAVEFORMS

CLOCK

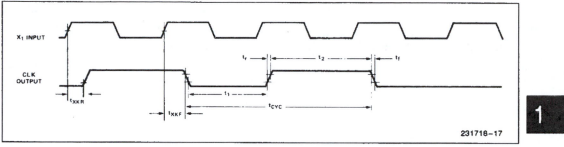

231718–17

READ

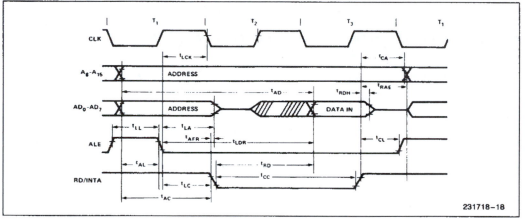

231718–18

WRITE

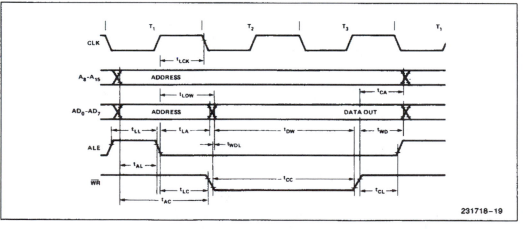

231718–19

8085AH/8085AH-2/8085AH-1

WAVEFORMS (Continued)

HOLD

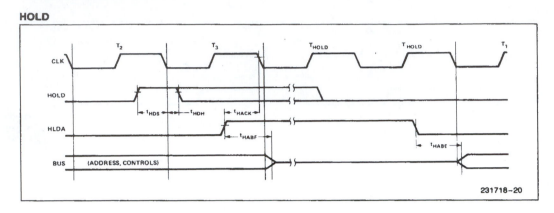

231718–20

READ OPERATION WITH WAIT CYCLE (TYPICAL)—SAME READY TIMING APPLIES TO WRITE

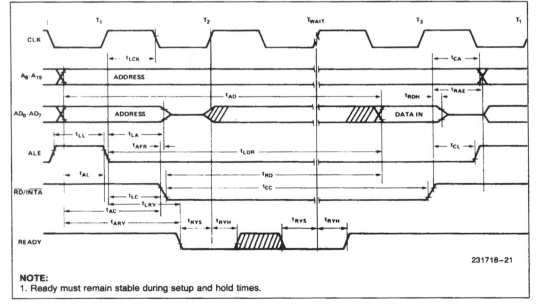

231718–21

NOTE:
1. Ready must remain stable during setup and hold times.

intel®   **8085AH/8085AH-2/8085AH-1**

WAVEFORMS (Continued)

INTERRUPT AND HOLD

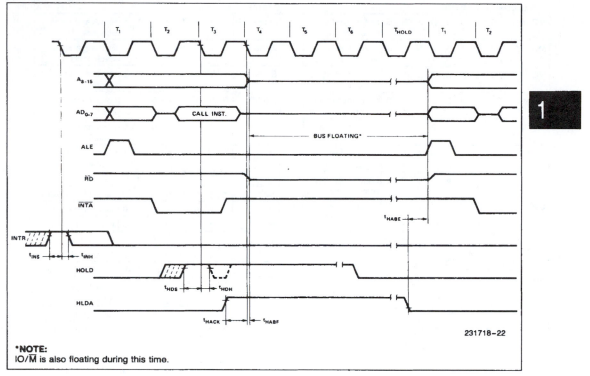

231718–22

*NOTE:
IO/M̄ is also floating during this time.

8085AH/8085AH-2/8085AH-1

Table 6. Instruction Set Summary

Mnemonic	D7	D6	D5	D4	D3	D2	D1	D0	Operations Description
MOVE, LOAD AND STORE									
MOVr1 r2	0	1	D	D	D	S	S	S	Move register to register
MOV M.r	0	1	1	1	0	S	S	S	Move register to memory
MOV r.M	0	1	D	D	D	1	1	0	Move memory to register
MVI r	0	0	D	D	D	1	1	0	Move immediate register
MVI M	0	0	1	1	0	1	1	0	Move immediate memory
LXI B	0	0	0	0	0	0	0	1	Load immediate register Pair B & C
LXI D	0	0	0	1	0	0	0	1	Load immediate register Pair D & E
LXI H	0	0	1	0	0	0	0	1	Load immediate register Pair H & L
STAX B	0	0	0	0	0	0	1	0	Store A indirect
STAX D	0	0	0	1	0	0	1	0	Store A indirect
LDAX B	0	0	1	0	1	0	1	0	Load A indirect
LDAX D	0	0	0	1	1	0	1	0	Load A indirect
STA	0	0	1	1	0	0	1	0	Store A direct
LDA	0	0	1	1	1	0	1	0	Load A direct
SHLD	0	0	1	0	0	0	1	0	Store H & L direct
LHLD	0	0	1	0	1	0	1	0	Load H & L direct
XCHG	1	1	1	0	1	0	1	1	Exchange D & E, H & L Registers
STACK OPS									
PUSH B	1	1	0	0	0	1	0	1	Push register Pair B & C on stack
PUSH D	1	1	0	1	0	1	0	1	Push register Pair D & E on stack
PUSH H	1	1	1	0	0	1	0	1	Push register Pair H & L on stack
PUSH PSW	1	1	1	1	0	1	0	1	Push A and Flags on stack
POP B	1	1	0	0	0	0	0	1	Pop register Pair B & C off stack
POP D	1	1	0	1	0	0	0	1	Pop register Pair D & E off stack
POP H	1	1	1	0	0	0	0	1	Pop register Pair H & L off stack

Mnemonic	D7	D6	D5	D4	D3	D2	D1	D0	Operations Description
STACK OPS (Continued)									
POP PSW	1	1	1	1	0	0	0	1	Pop A and Flags off stack
XTHL	1	1	1	0	0	0	1	1	Exchange top of stack, H & L
SPHL	1	1	1	1	1	0	0	1	H & L to stack pointer
LXI SP	0	0	1	1	0	0	0	1	Load immediate stack pointer
INX SP	0	0	1	1	0	0	1	1	Increment stack pointer
DCX SP	0	0	1	1	1	0	1	1	Decrement stack pointer
JUMP									
JMP	1	1	0	0	0	0	1	1	Jump unconditional
JC	1	1	0	1	1	0	1	0	Jump on carry
JNC	1	1	0	1	0	0	1	0	Jump on no carry
JZ	1	1	0	0	1	0	1	0	Jump on zero
JNZ	1	1	0	0	0	0	1	0	Jump on no zero
JP	1	1	1	1	0	0	1	0	Jump on positive
JM	1	1	1	1	1	0	1	0	Jump on minus
JPE	1	1	1	0	1	0	1	0	Jump on parity even
JPO	1	1	1	0	0	0	1	0	Jump on parity odd
PCHL	1	1	1	0	1	0	0	1	H & L to program counter
CALL									
CALL	1	1	0	0	1	1	0	1	Call unconditional
CC	1	1	0	1	1	1	0	0	Call on carry
CNC	1	1	0	1	0	1	0	0	Call on no carry
CZ	1	1	0	0	1	1	0	0	Call on zero
CNZ	1	1	0	0	0	1	0	0	Call on no zero
CP	1	1	1	1	0	1	0	0	Call on positive
CM	1	1	1	1	1	1	0	0	Call on minus
CPE	1	1	1	0	1	1	0	0	Call on parity even
CPO	1	1	1	0	0	1	0	0	Call on parity odd
RETURN									
RET	1	1	0	0	1	0	0	1	Return
RC	1	1	0	1	1	0	0	0	Return on carry
RNC	1	1	0	1	0	0	0	0	Return on no carry
RZ	1	1	0	0	1	0	0	0	Return on zero

intel. 8085AH/8085AH-2/8085AH-1

Table 6. Instruction Set Summary (Continued)

Mnemonic	D7	D6	D5	D4	D3	D2	D1	D0	Operations Description
RETURN (Continued)									
RNZ	1	1	0	0	0	0	0	0	Return on no zero
RP	1	1	1	1	0	0	0	0	Return on positive
RM	1	1	1	1	1	0	0	0	Return on minus
RPE	1	1	1	0	1	0	0	0	Return on parity even
RPO	1	1	1	0	0	0	0	0	Return on parity odd
RESTART									
RST	1	1	A	A	A	1	1	1	Restart
INPUT/OUTPUT									
IN	1	1	0	1	1	0	1	1	Input
OUT	1	1	0	1	0	0	1	1	Output
INCREMENT AND DECREMENT									
INR r	0	0	D	D	D	1	0	0	Increment register
DCR r	0	0	D	D	D	1	0	1	Decrement register
INR M	0	0	1	1	0	1	0	0	Increment memory
DCR M	0	0	1	1	0	1	0	1	Decrement memory
INX B	0	0	0	0	0	0	1	1	Increment B & C registers
INX D	0	0	0	1	0	0	1	1	Increment D & E registers
INX H	0	0	1	0	0	0	1	1	Increment H & L registers
DCX B	0	0	0	0	1	0	1	1	Decrement B & C
DCX D	0	0	0	1	1	0	1	1	Decrement D & E
DCX H	0	0	1	0	1	0	1	1	Decrement H & L
ADD									
ADD r	1	0	0	0	0	S	S	S	Add register to A
ADC r	1	0	0	0	1	S	S	S	Add register to A with carry
ADD M	1	0	C	0	0	1	1	0	Add memory to A
ADC M	1	0	0	0	1	1	1	0	Add memory to A with carry
ADI	1	1	0	0	0	1	1	0	Add immediate to A
ACI	1	1	0	0	1	1	1	0	Add immediate to A with carry
DAD B	0	0	0	0	1	0	0	1	Add B & C to H & L

Mnemonic	D7	D6	D5	D4	D3	D2	D1	D0	Operations Description
ADD (Continued)									
DAD D	0	0	0	1	1	0	0	1	Add D & E to H & L
DAD H	0	0	1	0	1	0	0	1	Add H & L to H & L
DAD SP	0	0	1	1	1	0	0	1	Add stack pointer to H & L
SUBTRACT									
SUB r	1	0	0	1	0	S	S	S	Subtract register from A
SBB r	1	0	0	1	1	S	S	S	Subtract register from A with borrow
SUB M	1	0	0	1	0	1	1	0	Subtract memory from A
SBB M	1	0	0	1	1	1	1	0	Subtract memory from A with borrow
SUI	1	1	0	1	0	1	1	0	Subtract immediate from A
SBI	1	1	0	1	1	1	1	0	Subtract immediate from A with borrow
LOGICAL									
ANA r	1	0	1	0	0	S	S	S	And register with A
XRA r	1	0	1	0	1	S	S	S	Exclusive OR register with A
ORA r	1	0	1	1	0	S	S	S	OR register with A
CMP r	1	0	1	1	1	S	S	S	Compare register with A
ANA M	1	0	1	0	0	1	1	0	And memory with A
XRA M	1	0	1	0	1	1	1	0	Exclusive OR memory with A
ORA M	1	0	1	1	0	1	1	0	OR memory with A
CMP M	1	0	1	1	1	1	1	0	Compare memory with A
ANI	1	1	1	0	0	1	1	0	And immediate with A
XRI	1	1	1	0	1	1	1	0	Exclusive OR immediate with A
ORI	1	1	1	1	0	1	1	0	OR immediate with A
CPI	1	1	1	1	1	1	1	0	Compare immediate with A

1

8085AH/8085AH-2/8085AH-1

intel®

Table 6. Instruction Set Summary (Continued)

Mnemonic	D_7	D_6	D_5	D_4	D_3	D_2	D_1	D_0	Operations Description
ROTATE									
RLC	0	0	0	0	0	1	1	1	Rotate A left
RRC	0	0	0	0	1	1	1	1	Rotate A right
RAL	0	0	0	1	0	1	1	1	Rotate A left through carry
RAR	0	0	0	1	1	1	1	1	Rotate A right through carry
SPECIALS									
CMA	0	0	1	0	1	1	1	1	Complement A
STC	0	0	1	1	0	1	1	1	Set carry
CMC	0	0	1	1	1	1	1	1	Complement carry
DAA	0	0	1	0	0	1	1	1	Decimal adjust A

Mnemonic	D_7	D_6	D_5	D_4	D_3	D_2	D_1	D_0	Operations Description
CONTROL									
EI	1	1	1	1	1	0	1	1	Enable Interrupts
DI	1	1	1	1	0	0	1	1	Disable Interrupt
NOP	0	0	0	0	0	0	0	0	No-operation
HLT	0	1	1	1	0	1	1	0	Halt
NEW 8085AH INSTRUCTIONS									
RIM	0	0	1	0	0	0	0	0	Read Interrupt Mask
SIM	0	0	1	1	0	0	0	0	Set Interrupt Mask

NOTES:
1. DDS or SSS: B 000, C 001, D 010, E011, H 100, L101, Memory 110, A 111.
2. Two possible cycle times (6/12) indicate instruction cycles dependent on condition flags.
*All mnemonics copyrighted © Intel Corporation 1976.

Appendix III—The 8051 Microcontroller Specification Sheet

INTEL CORP (UP/PRPHLS)   20E **D** ■ 4826175 0079078 8 ■

intel®

PRELIMINARY

T-49-19-07

MCS®-51
8-BIT CONTROL-ORIENTED MICROCOMPUTERS
8031/8051
8031AH/8051AH
8032AH/8052AH
8751H/8751H-8

- ■ High Performance HMOS Process
- ■ Internal Timers/Event Counters
- ■ 2-Level Interrupt Priority Structure
- ■ 32 I/O Lines (Four 8-Bit Ports)
- ■ 64K Program Memory Space
- ■ Security Feature Protects EPROM Parts Against Software Piracy

- ■ Boolean Processor
- ■ Bit-Addressable RAM
- ■ Programmable Full Duplex Serial Channel
- ■ 111 Instructions (64 Single-Cycle)
- ■ 64K Data Memory Space

The MCS®-51 products are optimized for control applications. Byte-processing and numerical operations on small data structures are facilitated by a variety of fast addressing modes for accessing the internal RAM. The instruction set provides a convenient menu of 8-bit arithmetic instructions, including multiply and divide instructions. Extensive on-chip support is provided for one-bit variables as a separate data type, allowing direct bit manipulation and testing in control and logic systems that require Boolean processing.

The 8051 is the original member of the MCS-51 family. The 8051AH is identical to the 8051, but it is fabricated with HMOS II technology.

The 8751H is an EPROM version of the 8051AH; that is, the on-chip Program Memory can be electrically programmed, and can be erased by exposure to ultraviolet light. It is fully compatible with its predecessor, the 8751-8, but incorporates two new features: a Program Memory Security bit that can be used to protect the EPROM against unauthorized read-out, and a programmable baud rate modification bit (SMOD). The 8751H-8 is identical to the 8751H but only operates up to 8 MHz.

The 8052AH is an enhanced version of the 8051AH. It is backwards compatible with the 8051AH and is fabricated with HMOS II technology. The 8052AH enhancements are listed in the table below. Also refer to this table for the ROM, ROMless, and EPROM versions of each product.

Device	Internal Memory		Timers/ Event Counters	Interrupts
	Program	**Data**		
8052AH	8K x 8 ROM	256 x 8 RAM	3 x 16-Bit	6
8051AH	4K x 8 ROM	128 x 8 RAM	2 x 16-Bit	5
8051	4K x 8 ROM	128 x 8 RAM	2 x 16-Bit	5
8032AH	none	256 x 8 RAM	3 x 16-Bit	6
8031AH	none	128 x 8 RAM	2 x 16-Bit	5
8031	none	128 x 8 RAM	2 x 16-Bit	5
8751H	4K x 8 EPROM	128 x 8 RAM	2 x 16-Bit	5
8751H-8	4K x 8 EPROM	128 x 8 RAM	2 x 16-Bit	5

October 1988
Order Number: 270048-004

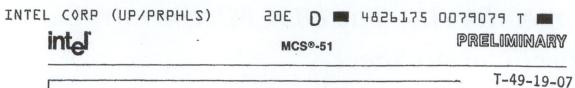

MCS®-51   PRELIMINARY

T-49-19-07

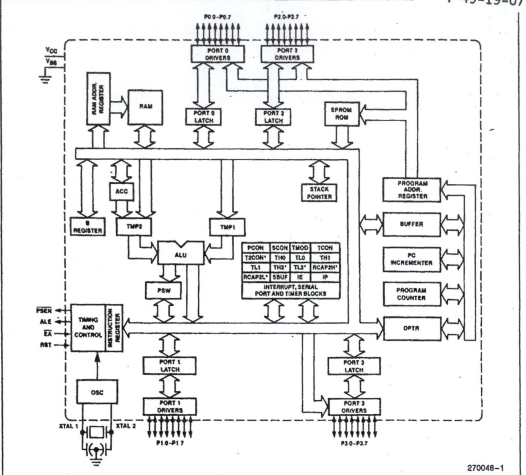

Figure 1. MCS®-51 Block Diagram

PACKAGES

Part	Prefix	Package Type
8051AH/	P	40-Pin Plastic DIP
8031AH	D	40-Pin CERDIP
	N	44-Pin PLCC
8052AH/	P	40-Pin Plastic DIP
8032AH	D	40-Pin CERDIP
	N	44-Pin PLCC
8751H/	D	40-Pin CERDIP
8751H-8	R	44-Pin LCC

PIN DESCRIPTIONS

V_{CC}: Supply voltage.

V_{SS}: Circuit ground.

Port 0: Port 0 is an 8-bit open drain bidirectional I/O port. As an output port each pin can sink 8 LS TTL inputs.

Port 0 pins that have 1s written to them float, and in that state can be used as high-impedance inputs.

Port 0 is also the multiplexed low-order address and data bus during accesses to external Program and Data Memory. In this application it uses strong internal pullups when emitting 1s and can source and sink 8 LS TTL inputs.

Port 0 also receives the code bytes during programming of the EPROM parts, and outputs the code bytes during program verification of the ROM and EPROM parts. External pullups are required during program verification.

INTEL CORP (UP/PRPHLS)   20E **D** ■ 4826175 0079080 6 ■

intel   MCS®-51   PRELIMINARY

T-49-19-07

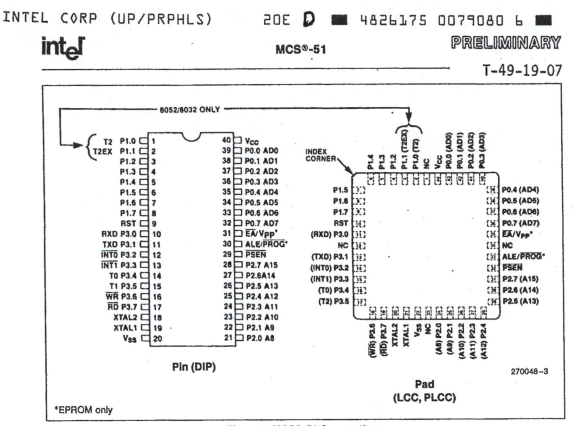

Figure 2. MCS®-51 Connections

Port 1: Port 1 is an 8-bit bidirectional I/O port with internal pullups. The Port 1 output buffers can sink/source 4 LS TTL inputs. Port 1 pins that have 1s written to them are pulled high by the internal pullups, and in that state can be used as inputs. As inputs, Port 1 pins that are externally being pulled low will source current (I_{IL} on the data sheet) because of the internal pullups.

Port 1 also receives the low-order address bytes during programming of the EPROM parts and during program verification of the ROM and EPROM parts.

In the 8032AH and 8052AH, Port 1 pins P1.0 and P1.1 also serve the T2 and T2EX functions, respectively.

Port 2: Port 2 is an 8-bit bidirectional I/O port with internal pullups. The Port 2 output buffers can sink/source 4 LS TTL inputs. Port 2 pins that have 1s written to them are pulled high by the internal pullups, and in that state can be used as inputs. As inputs, Port 2 pins that are externally being pulled low will source current (I_{IL} on the data sheet) because of the internal pullups.

Port 2 emits the high-order address byte during fetches from external Program Memory and during accesses to external Data Memory that use 16-bit addresses (MOVX @DPTR). In this application it uses strong internal pullups when emitting 1s. During accesses to external Data Memory that use 8-bit addresses (MOVX @Ri), Port 2 emits the contents of the P2 Special Function Register.

Port 2 also receives the high-order address bits during programming of the EPROM parts and during program verification of the ROM and EPROM parts.

Port 3: Port 3 is an 8-bit bidirectional I/O port with internal pullups. The Port 3 output buffers can sink/source 4 LS TTL inputs. Port 3 pins that have 1s written to them are pulled high by the internal pullups, and in that state can be used as inputs. As inputs, Port 3 pins that are externally being pulled low will source current (I_{IL} on the data sheet) because of the pullups.

Port 3 also serves the functions of various special features of the MCS-51 Family, as listed below:

Port Pin	Alternative Function
P3.0	RXD (serial input port)
P3.1	TXD (serial output port)
P3.2	INT0 (external interrupt 0)
P3.3	INT1 (external interrupt 1)
P3.4	T0 (Timer 0 external input)
P3.5	T1 (Timer 1 external input)
P3.6	WR (external data memory write strobe)
P3.7	RD (external data memory read strobe)

INTEL CORP (UP/PRPHLS)   20E D ■ 4826175 0079081 8 ■

intel   MCS®-51   PRELIMINARY

T-49-19-07

RST: Reset input. A high on this pin for two machine cycles while the oscillator is running resets the device.

ALE/PROG: Address Latch Enable output pulse for latching the low byte of the address during accesses to external memory. This pin is also the program pulse input (PROG) during programming of the EPROM parts.

In normal operation ALE is emitted at a constant rate of ⅙ the oscillator frequency, and may be used for external timing or clocking purposes. Note, however, that one ALE pulse is skipped during each access to external Data Memory.

PSEN: Program Store Enable is the read strobe to external Program Memory.

When the device is executing code from external Program Memory, PSEN is activated twice each machine cycle, except that two PSEN activations are skipped during each access to external Data Memory.

EA/V$_{PP}$: External Access enable EA must be strapped to V$_{SS}$ in order to enable any MCS-51 device to fetch code from external Program memory locations starting at 0000H up to FFFFH. EA must be strapped to V$_{CC}$ for internal program execution.

Note, however, that if the Security Bit in the EPROM devices is programmed, the device will not fetch code from any location in external Program Memory.

This pin also receives the 21V programming supply voltage (VPP) during programming of the EPROM parts.

XTAL1: Input to the inverting oscillator amplifier.

XTAL2: Output from the inverting oscillator amplifier.

OSCILLATOR CHARACTERISTICS

XTAL1 and XTAL2 are the input and output, respectively, of an inverting amplifier which can be configured for use as an on-chip oscillator, as shown in Figure 3. Either a quartz crystal or ceramic resonator may be used. More detailed information concerning the use of the on-chip oscillator is available in Application Note AP-155, "Oscillators for Microcontrollers."

To drive the device from an external clock source, XTAL1 should be grounded, while XTAL2 is driven, as shown in Figure 4. There are no requirements on the duty cycle of the external clock signal, since the input to the internal clocking circuitry is through a divide-by-two flip-flop, but minimum and maximum high and low times specified on the Data Sheet must be observed.

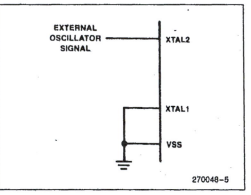

Figure 4. External Drive Configuration

DESIGN CONSIDERATIONS

If an 8751BH or 8752BH may replace an 8751H in a future design, the user should carefully compare both data sheets for DC or AC Characteristic differences. Note that the V$_{IH}$ and I$_{IH}$ specifications for the EA pin differ significantly between the devices.

Exposure to light when the EPROM device is in operation may cause logic errors. For this reason, it is suggested that an opaque label be placed over the window when the die is exposed to ambient light.

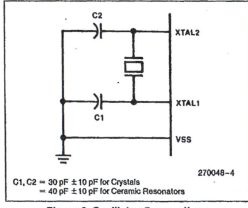

C1, C2 = 30 pF ± 10 pF for Crystals
= 40 pF ± 10 pF for Ceramic Resonators

270048-4

Figure 3. Oscillator Connections

INTEL CORP (UP/PRPHLS)   20E **D** ■ 4826175 0079082 T ■

intel   MCS®-51   **PRELIMINARY**

T-49-19-07

ABSOLUTE MAXIMUM RATINGS*

Ambient Temperature Under Bias ......0°C to 70°C

Storage Temperature ....−65°C to +150°C

Voltage on EA/V$_{PP}$ Pin to V$_{SS}$ ... −0.5V to +21.5V

Voltage on Any Other Pin to V$_{SS}$ .... −0.5V to +7V

Power Dissipation..........................1.5W

*Notice: Stresses above those listed under "Absolute Maximum Ratings" may cause permanent damage to the device. This is a stress rating only and functional operation of the device at these or any other conditions above those indicated in the operational sections of this specification is not implied. Exposure to absolute maximum rating conditions for extended periods may affect device reliability.

D.C. CHARACTERISTICS T$_A$ = 0°C to 70°C; V$_{CC}$ = 5V ±10%; V$_{SS}$ = 0V

Symbol	Parameter		Min	Max	Units	Test Conditions
V$_{IL}$	Input Low Voltage (Except EA Pin of 8751H & 8751H-8)		−0.5	0.8	V	
V$_{IL1}$	Input Low Voltage to EA Pin of 8751H & 8751H-8		0	0.7	V	
V$_{IH}$	Input High Voltage (Except XTAL2, RST)		2.0	V$_{CC}$ + 0.5	V	
V$_{IH1}$	Input High Voltage to XTAL2, RST		2.5	V$_{CC}$ + 0.5	V	XTAL1 = V$_{SS}$
V$_{OL}$	Output Low Voltage (Ports 1, 2, 3)*			0.45	V	I$_{OL}$ = 1.6 mA
V$_{OL1}$	Output Low Voltage (Port 0, ALE, PSEN)*					
		8751H, 8751H-8		0.60	V	I$_{OL}$ = 3.2 mA
				0.45	V	I$_{OL}$ = 2.4 mA
		All Others		0.45	V	I$_{OL}$ = 3.2 mA
V$_{OH}$	Output High Voltage (Ports 1, 2, 3, ALE, PSEN)		2.4		V	I$_{OH}$ = −80 µA
V$_{OH1}$	Output High Voltage (Port 0 in External Bus Mode)		2.4		V	I$_{OH}$ = −400 µA
I$_{IL}$	Logical 0 Input Current (Ports 1, 2, 3, RST) 8032AH, 8052AH			−800	µA	V$_{IN}$ = 0.45V
	All Others			−500	µA	V$_{IN}$ = 0.45V
I$_{IL1}$	Logical 0 Input Current to EA Pin of 8751H & 8751H-8 Only			−15	mA	V$_{IN}$ = 0.45V
I$_{IL2}$	Logical 0 Input Current (XTAL2)			−3.2	mA	V$_{IN}$ = 0.45V
I$_{LI}$	Input Leakage Current (Port 0) 8751H & 8751H-8			±100	µA	0.45 ≤ V$_{IN}$ ≤ V$_{CC}$
	All Others			±10	µA	0.45 ≤ V$_{IN}$ ≤ V$_{CC}$
I$_{IH}$	Logical 1 Input Current to EA Pin of 8751H & 8751H-8			500	µA	V$_{IN}$ = 2.4V
I$_{IH1}$	Input Current to RST to Activate Reset			500	µA	V$_{IN}$ < (V$_{CC}$ − 1.5V)
I$_{CC}$	Power Supply Current: 8031/8051			160	mA	
	8031AH/8051AH			125	mA	All Outputs
	8032AH/8052AH			175	mA	Disconnected;
	8751H/8751H-8			250	mA	EA = V$_{CC}$
C$_{IO}$	Pin Capacitance			10	pF	Test freq = 1 MHz

*NOTE:

Capacitive loading on Ports 0 and 2 may cause spurious noise pulses to be superimposed on the V$_{OL}$s of ALE and Ports 1 and 3. The noise is due to external bus capacitance discharging into the Port 0 and Port 2 pins when these pins make 1-to-0 transitions during bus operations. In the worst cases (capacitive loading > 100 pF), the noise pulse on the ALE line may exceed 0.8V. In such cases it may be desirable to qualify ALE with a Schmitt Trigger, or use an address latch with a Schmitt Trigger STROBE input.

intel   MCS®-51   PRELIMINARY

T-49-19-07

A.C. CHARACTERISTICS $T_A = 0°C$ to $+70°C$; $V_{CC} = 5V \pm 10\%$; $V_{SS} = 0V$;
Load Capacitance for Port 0, ALE, and PSEN = 100 pF;
Load Capacitance for All Other Outputs = 80 pF

Symbol	Parameter	12 MHz Oscillator		Variable Oscillator		Units
		Min	Max	Min	Max	
1/TCLCL	Oscillator Frequency			3.5	12.0	MHz
TLHLL	ALE Pulse Width	127		2TCLCL−40		ns
TAVLL	Address Valid to ALE Low	43		TCLCL−40		ns
TLLAX	Address Hold after ALE Low	48		TCLCL−35		ns
TLLIV	ALE Low to Valid Instr In					
	8751H		183		4TCLCL−150	ns
	All Others		233		4TCLCL−100	ns
TLLPL	ALE Low to PSEN Low	58		TCLCL−25		ns
TPLPH	PSEN Pulse Width					
	8751H	190		3TCLCL−60		ns
	All Others	215		3TCLCL−35		ns
TPLIV	PSEN Low to Valid Instr In					
	8751H		100		3TCLCL−150	ns
	All Others		125		3TCLCL−125	ns
TPXIX	Input Instr Hold after PSEN	0		0		ns
TPXIZ	Input Instr Float after PSEN		63		TCLCL−20	ns
TPXAV	PSEN to Address Valid	75		TCLCL−8		ns
TAVIV	Address to Valid Instr In					
	8751H		267		5TCLCL−150	ns
	All Others		302		5TCLCL−115	ns
TPLAZ	PSEN Low to Address Float		20		20	ns
TRLRH	RD Pulse Width	400		6TCLCL−100		ns
TWLWH	WR Pulse Width	400		6TCLCL−100		ns
TRLDV	RD Low to Valid Data In		252		5TCLCL−165	ns
TRHDX	Data Hold after RD	0		0		ns
TRHDZ	Data Float after RD		97		2TCLCL−70	ns
TLLDV	ALE Low to Valid Data In		517		8TCLCL−150	ns
TAVDV	Address to Valid Data In		585		9TCLCL−165	ns
TLLWL	ALE Low to RD or WR Low	200	300	3TCLCL−50	3TCLCL+50	ns
TAVWL	Address to RD or WR Low	203		4TCLCL−130		ns
TQVWX	Data Valid to WR Transition					
	8751H	13		TCLCL−70		ns
	All Others	23		TCLCL−60		ns
TQVWH	Data Valid to WR High	433		7TCLCL−150		ns
TWHQX	Data Hold after WR	33		TCLCL−50		ns
TRLAZ	RD Low to Address Float		20		20	ns
TWHLH	RD or WR High to ALE High					
	8751H	33	133	TCLCL−50	TCLCL+50	ns
	All Others	43	123	TCLCL−40	TCLCL+40	ns

NOTE:
*This table does not include the 8751-8 A.C. characteristics (see next page).

intel   MCS®-51   PRELIMINARY

T-49-19-07

This Table is only for the 8751H-8

A.C. CHARACTERISTICS $T_A = 0°C$ to $+70°C$; $V_{CC} = 5V \pm 10\%$; $V_{SS} = 0V$;
Load Capacitance for Port 0, ALE, and PSEN = 100 pF;
Load Capacitance for All Other Outputs = 80 pF

Symbol	Parameter	8 MHz Oscillator		Variable Oscillator		Units
		Min	Max	Min	Max	
1/TCLCL	Oscillator Frequency			3.5	8.0	MHz
TLHLL	ALE Pulse Width	210		2TCLCL−40		ns
TAVLL	Address Valid to ALE Low	85		TCLCL−40		ns
TLLAX	Address Hold after ALE Low	90		TCLCL−35		ns
TLLIV	ALE Low to Valid Instr In		350		4TCLCL−150	ns
TLLPL	ALE Low to PSEN Low	100		TCLCL−25		ns
TPLPH	PSEN Pulse Width	315		3TCLCL−60		ns
TPLIV	PSEN Low to Valid Instr In		225		3TCLCL−150	ns
TPXIX	Input Instr Hold after PSEN	0		0		ns
TPXIZ	Input Instr Float after PSEN		105		TCLCL−20	ns
TPXAV	PSEN to Address Valid	117		TCLCL−8		ns
TAVIV	Address to Valid Instr In		475		5TCLCL−150	ns
TPLAZ	PSEN Low to Address Float		20		20	ns
TRLRH	RD Pulse Width	650		6TCLCL−100		ns
TWLWH	WR Pulse Width	650		6TCLCL−100		ns
TRLDV	RD Low to Valid Data In		460		5TCLCL−165	ns
TRHDX	Data Hold after RD	0		0		ns
TRHDZ	Data Float after RD		180		2TCLCL−70	ns
TLLDV	ALE Low to Valid Data In		850		8TCLCL−150	ns
TAVDV	Address to Valid Data In		960		9TCLCL−165	ns
TLLWL	ALE Low to RD or WR Low	325	425	3TCLCL−50	3TCLCL+50	ns
TAVWL	Address to RD or WR Low	370		4TCLCL−130		ns
TQVWX	Data Valid to WR Transition	55		TCLCL−70		ns
TQVWH	Data Valid to WR High	725		7TCLCL−150		ns
TWHQX	Data Hold after WR	75		TCLCL−50		ns
TRLAZ	RD Low to Address Float		20		20	ns
TWHLH	RD or WR High to ALE High	75	175	TCLCL−50	TCLCL+50	ns

INTEL CORP (UP/PRPHLS)   20E D ■ 4826175 0079085 5 ■

intel   MCS®-51   **PRELIMINARY**

EXTERNAL PROGRAM MEMORY READ CYCLE

T-49-19-07

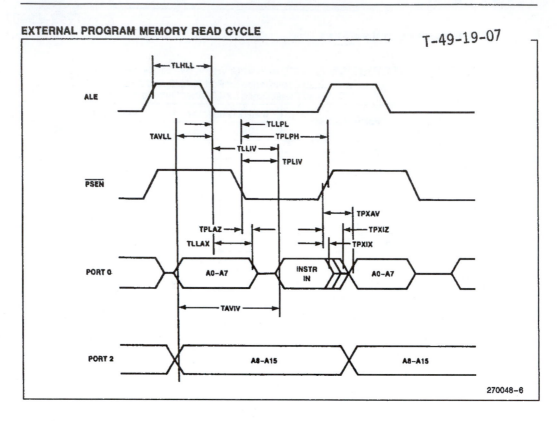

270048-6

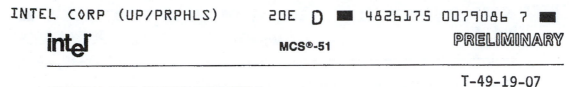

intel   MCS®-51   **PRELIMINARY**

T-49-19-07

EXTERNAL DATA MEMORY READ CYCLE

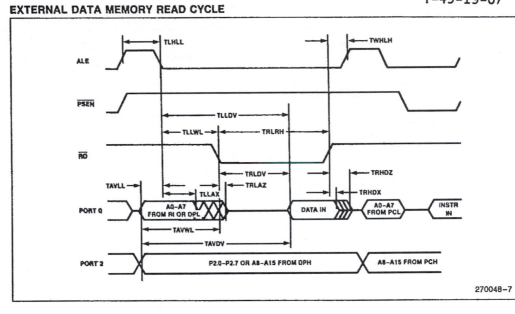

270048–7

EXTERNAL DATA MEMORY WRITE CYCLE

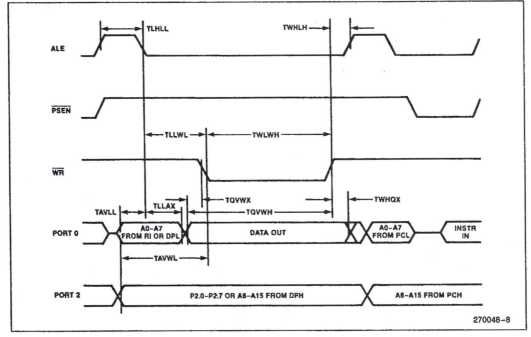

270048–8

INTEL CORP (UP/PRPHLS) 20E D ■ 4826175 0079087 9 ■

intel

MCS®-51

T-49-19-07

SERIAL PORT TIMING—SHIFT REGISTER MODE

Test Conditions: T_A = 0°C to 70°C; VCC = 5V ±10%; VSS = 0V; Load Capacitance = 80 pF

Symbol	Parameter	12 MHz Oscillator		Variable Oscillator		Units
		Min	Max	Min	Max	
TXLXL	Serial Port Clock Cycle Time	1.0		12TCLCL		μS
TQVXH	Output Data Setup to Clock Rising Edge	700		10TCLCL − 133		ns
TXHQX	Output Data Hold after Clock Rising Edge	50		2TCLCL − 117		ns
TXHDX	Input Data Hold after Clock Rising Edge	0		0		ns
TXHDV	Clock Rising Edge to Input Data Valid		700		10TCLCL − 133	ns

SHIFT REGISTER TIMING WAVEFORMS

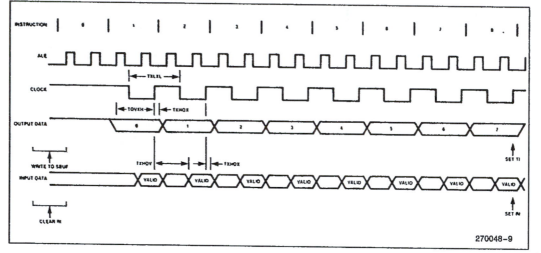

270048–9

INTEL CORP (UP/PRPHLS)   20E **D**   ■   4826175 0079088 0 ■

int_el   MCS®-51   PRELIMINARY

T-49-19-07

EXTERNAL CLOCK DRIVE

Symbol	Parameter	Min	Max	Units
1/TCLCL	Oscillator Frequency (except 8751H-8)	3.5	12	MHz
	8751H-8	3.5	8	MHz
TCHCX	High Time	20		ns
TCLCX	Low Time	20		ns
TCLCH	Rise Time		20	ns
TCHCL	Fall Time		20	ns

EXTERNAL CLOCK DRIVE WAVEFORM

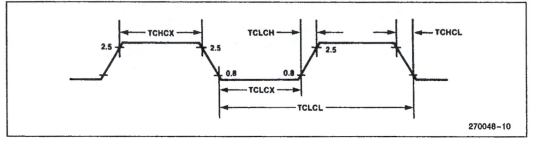

270048-10

A.C. TESTING INPUT, OUTPUT WAVEFORM

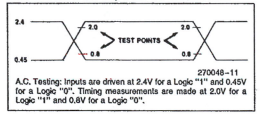

270048-11

A.C. Testing: Inputs are driven at 2.4V for a Logic "1" and 0.45V for a Logic "0". Timing measurements are made at 2.0V for a Logic "1" and 0.8V for a Logic "0".

intel   MCS®-51   **PRELIMINARY**

EPROM CHARACTERISTICS

Table 3. EPROM Programming Modes

Mode	RST	$\overline{PSEN}$	ALE	$\overline{EA}$	P2.7	P2.6	P2.5	P2.4
Program	1	0	0*	VPP	1	0	X	X
Inhibit	1	0	1	X	1	0	X	X
Verify	1	0	1	1	0	0	X	X
Security Set	1	0	0*	VPP	1	1	X	X

NOTE:
"1" = logic high for that pin
"0" = logic low for that pin
"X" = "don't care"

"VPP" = +21V ±0.5V
*ALE is pulsed low for 50 ms.

Programming the EPROM

To be programmed, the part must be running with a 4 to 6 MHz oscillator. (The reason the oscillator needs to be running is that the internal bus is being used to transfer address and program data to appropriate internal registers.) The address of an EPROM location to be programmed is applied to Port 1 and pins P2.0–P2.3 of Port 2, while the code byte to be programmed into that location is applied to Port 0. The other Port 2 pins, and RST, $\overline{PSEN}$, and $\overline{EA}$ should be held at the "Program" levels indicated in Table 3. ALE is **pulsed** low for 50 ms to program the code byte into the addressed EPROM location. The setup is shown in Figure 5.

Normally $\overline{EA}$ is held at a logic high until just before ALE is to be pulsed. Then $\overline{EA}$ is raised to +21V, ALE is pulsed, and then $\overline{EA}$ is returned to a logic high. Waveforms and detailed timing specifications are shown in later sections of this data sheet.

Note that the $\overline{EA}$/VPP pin must not be allowed to go above the maximum specified VPP level of 21.5V for any amount of time. Even a narrow glitch above that voltage level can cause permanent damage to the device. The VPP source should be well regulated and free of glitches.

Program Verification

If the Security Bit has not been programmed, the on-chip Program Memory can be read out for verification purposes, if desired, either during or after the programming operation. The address of the Program Memory location to be read is applied to Port 1 and pins P2.0–P2.3. The other pins should be held at the "Verify" levels indicated in Table 3. The contents of the addressed location will come out on Port 0. External pullups are required on Port 0 for this operation.

The setup, which is shown in Figure 6, is the same as for programming the EPROM except that pin P2.7 is held at a logic low, or may be used as an active-low read strobe.

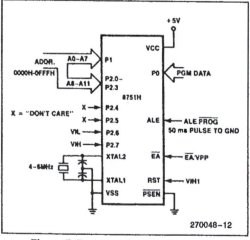

Figure 5. Programming Configuration

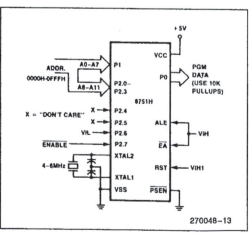

Figure 6. Program Verification

intel   MCS®-51   **PRELIMINARY**

T-49-19-07

EPROM Security

The security feature consists of a "locking" bit which when programmed denies electrical access by any external means to the on-chip Program Memory. The bit is programmed as shown in Figure 7. The setup and procedure are the same as for normal EPROM programming, except that P2.6 is held at a logic high. Port 0, Port 1, and pins P2.0–P2.3 may be in any state. The other pins should be held at the "Security" levels indicated in Table 3.

Once the Security Bit has been programmed, it can be cleared only by full erasure of the Program Memory. While it is programmed, the internal Program Memory can not be read out, the device can not be further programmed, and it **can not execute out of external program memory.** Erasing the EPROM, thus clearing the Security Bit, restores the device's full functionality. It can then be reprogrammed.

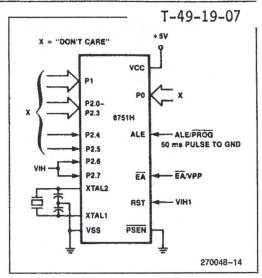

Figure 7. Programming the Security Bit

Erasure Characteristics

Erasure of the EPROM begins to occur when the chip is exposed to light with wavelengths shorter than approximately 4,000 Angstroms. Since sunlight and fluorescent lighting have wavelengths in this range, exposure to these light sources over an extended time (about 1 week in sunlight, or 3 years in room-level fluorescent lighting) could cause inadvertent erasure. If an application subjects the device to this type of exposure, it is suggested that an opaque label be placed over the window.

The recommended erasure procedure is exposure to ultraviolet light (at 2537 Angstroms) to an integrated dose of at least 15 W-sec/cm^2. Exposing the EPROM to an ultraviolet lamp of 12,000 μW/cm^2 rating for 20 to 30 minutes, at a distance of about 1 inch, should be sufficient.

Erasure leaves the array in an all 1s state.

EPROM PROGRAMMING AND VERIFICATION CHARACTERISTICS

T_A = 21°C to 27°C; VCC = 5V $\pm$10%; VSS = 0V

Symbol	Parameter	Min	Max	Units
VPP	Programming Supply Voltage	20.5	21.5	V
IPP	Programming Supply Current		30	mA
1/TCLCL	Oscillator Frequency	4	6	MHz
TAVGL	Address Setup to $\overline{PROG}$ Low	48TCLCL		
TGHAX	Address Hold after $\overline{PROG}$	48TCLCL		
TDVGL	Data Setup to $\overline{PROG}$ Low	48TCLCL		
TGHDX	Data Hold after $\overline{PROG}$	48TCLCL		
TEHSH	P2.7 ($\overline{ENABLE}$) High to VPP	48TCLCL		
TSHGL	VPP Setup to $\overline{PROG}$ Low	10		μs
TGHSL	VPP Hold after $\overline{PROG}$	10		μs
TGLGH	$\overline{PROG}$ Width	45	55	ms
TAVQV	Address to Data Valid		48TCLCL	
TELQV	$\overline{ENABLE}$ Low to Data Valid		48TCLCL	
TEHQZ	Data Float after $\overline{ENABLE}$	0	48TCLCL	

MCS®-51   PRELIMINARY

T-49-19-07

EPROM PROGRAMMING AND VERIFICATION WAVEFORMS

For programming conditions see Figure 5.   For verification conditions see Figure 6.   270048–15

DATA SHEET REVISION SUMMARY

The following are the key differences between this and the -003 version of this data sheet:

1. Introduction was expanded to include product descriptions.
2. Package table was added.
3. Design Considerations added.
4. Test Conditions for I_{IL1} and I_{IH} specifications added to the DC Characteristics.
5. Data Sheet Revision Summary added.

Index